Alcohol at Work

From uninsured losses to workplace injuries, from absenteeism to addiction, alcohol consumption can have a major impact on the workplace. Lost productivity alone costs the UK economy more than £7 billion annually. What's more, the impact of alcohol abuse on the employee and those around them is a wellbeing issue that all employers need to be alert to, especially with many now choosing to work from home.

Alcohol at Work: Tackling Workplace Drinking to Improve Wellbeing will guide health and safety professionals through the practical steps that they can take to mitigate risk and improve wellbeing, safeguarding and inclusivity. The book explores key topics such as driving legislation, the use of alcohol testing, stressful jobs and designing and implementing an alcohol policy that uses practical examples that highlight common workplace scenarios. It translates evidence-based research into workable solutions using tools that enable organisations to influence workplace culture, policies and practices so they can meet their legal obligations and ensure a thriving workforce. It delves into the legal aspects of alcohol abuse at work, offering treatment and adopting an alcohol and wellbeing policy for a business. This comprehensive guide will help the reader to understand and adopt practices that mitigate the risks of employees developing an alcohol problem and ensure that employers can tackle the issue head on should it arise.

This book is an essential read for any professional in occupational health and safety, human resources, business and management, policy creation, insurance and corporate governance.

Lauren Booker has worked for over twenty years in the alcohol sector and specialises in supporting the private sector to reduce alcohol-related harm and promote employee wellbeing. She also regularly advises firms on the development and implementation of effective policies and practices around alcohol and the workplace. She is a qualified alcohol treatment practitioner and has extensive experience as an alcohol counsellor and coach. Lauren has advised the UK government on the development of a model alcohol policy for employers, is an experienced speaker at national wellbeing events and runs a blog on alcohol.

Alcohol at Work

Tackling Workplace Drinking to Improve Wellbeing

Lauren Booker

CRC Press
Taylor & Francis Group
Boca Raton London New York

CRC Press is an imprint of the
Taylor & Francis Group, an **informa** business

Designed cover image: Shutterstock

First edition published 2026
by CRC Press
2385 NW Executive Center Drive, Suite 320, Boca Raton FL 33431

and by CRC Press
4 Park Square, Milton Park, Abingdon, Oxon, OX14 4RN

CRC Press is an imprint of Taylor & Francis Group, LLC

Library of Congress Cataloging-in-Publication Data
Names: Booker, Lauren author
Title: Alcohol at work : tackling workplace drinking to improve wellbeing / Lauren Booker.
Description: First edition. | Boca Raton, FL : CRC Press, 2025. | Includes bibliographical references and index.
Identifiers: LCCN 2025005962 (print) | LCCN 2025005963 (ebook) | ISBN 9781032645988 hardback | ISBN 9781032648910 paperback | ISBN 9781032648927 ebook
Subjects: LCSH: Alcoholism and employment | Employees—Alcohol use | Problem employees
Classification: LCC HF5549.5.A4 B66 2025 (print) | LCC HF5549.5.A4 (ebook) | DDC 658.3/822—dc23/eng/20250320
LC record available at https://lccn.loc.gov/2025005962
LC ebook record available at https://lccn.loc.gov/2025005963

ISBN: 978-1-032-64598-8 (hbk)
ISBN: 978-1-032-64891-0 (pbk)
ISBN: 978-1-032-64892-7 (ebk)

DOI: 10.1201/9781032648927

Typeset in Times
by Apex CoVantage, LLC

Access the Support Material: Routledge.com/9781032645988

Contents

Introduction

In workplaces throughout the world, a quiet epidemic is brewing. It's an epidemic that rarely makes headlines, and yet it exerts its influence over countless workplaces, affecting the lives of employees and the health of organisations. This epidemic is the prevalence of problematic alcohol use, and it's time to shine a spotlight on it.

From a health perspective, alcohol is a dangerous substance, contributing to the development of cancers, harming DNA and proteins, impairing cognitive function, coordination and memory and fuelling mental health difficulties. It harms the foetus in utero and causes dementia in adults. In fact, it's a causal factor in over 200 health conditions.[1] Drinking, though, is also one of our favourite pastimes as a species. In the USA, there were over 108,000 alcohol-related deaths in 2021[2] alone, and over 3 million deaths worldwide.[1]

Here in the UK, Scotland recorded a fourteen-year high of alcohol-related deaths in 2022.[3]

Alcohol also disrupts the balance of neurotransmitters in the brain and over time, contributes to alcohol-induced dementia. It can weaken the immune system, making the body more susceptible to infections. It impairs the function of cells and their ability to respond effectively to pathogens, leaving the body less able to defend itself against diseases and is also linked to the development of cancers.

And that's just the physical impact. From family breakdown to violent crime, from drink-driving to child protection, excessive alcohol consumption lays a heavy burden at the door of our social care structures. Almost one in three adults in the UK is affected by a loved one's drinking and over 1.5 million children live in a household with a high-risk drinker.[4] The consequences of our alco-centric society reverberate throughout our society and impact all of us.

For the concerned employer, alcohol is a major cause of lost productivity, absenteeism and accidents, leading to a staggering £7.3 billion annual cost to commerce and industry.[5] This book is scattered with cautionary tales of the devastation that can occur when alcohol and employment collide. Knowing how things can go wrong can help you get it right in your workplace. Despite this rather bleak picture, there's a lot that you can do to reduce the impact of alcohol on your workforce and this book explains what and how. Distilling key learning from two decades of working with employers and employees in a range of industries, the aim is to provide a set of clear, actionable steps for any employer, manager, occupational health (OH) or human resources (HR) professional to follow. Whether you're responding to existing concerns, overhauling your current policies and practices, initiating a preventative strategy or want to broaden your workplace wellbeing offer, the pages that follow will guide you through the pitfalls and opportunities that arise.

You have probably already observed that alcohol harm in the workplace is a complex issue with far-reaching consequences and alcohol problems can be shaped by the nature of the work, the environment and employment practices. As HR and OH managers, we

DOI: 10.1201/9781032648927-1

need to be able to recognise the signs, understand the risk factors and take proactive steps to address this problem. In this book, we will examine the roots of this hidden epidemic, discussing its prevalence, potential impact and the risks that contribute to it.

We will also explore some of the solutions.

This book is not designed as a cover-to-cover read; rather, it's a dip-in-and-out guide to reducing alcohol harms. Each chapter contains checklists, questions for personal reflection and ideas on how you can protect your organisation and your colleagues from risk. It is intended to act as a critical friend. You'll find some stats and figures in the following pages, but not too many. After all, if you weren't already convinced that alcohol can impact your workplace, you wouldn't be reading this book.

We will explore strategies and best practices for creating a workplace that promotes wellbeing and addresses alcohol misuse head-on. We will hear from experts, share real-life stories and provide actionable guidance for HR and OH professionals committed to changing the trajectory of this hidden epidemic.

The journey ahead will not be easy, but it is essential. Together, we can transform workplace drinking culture, creating and safer environments for all employees. It's time to shine a light on alcohol and take meaningful steps towards change. In these chapters, we will identify evidence-based interventions that managers can implement to address problems. We will examine the role of policies, education and support systems in creating more responsible attitudes within organisations.

Additionally, we will explore the importance of leadership in setting the tone for alcohol use (or not) in the workplace. Effective leaders can model responsible behaviour and create an atmosphere where employees feel safe seeking help when needed. We will also discuss the role of **Employee Assistance Programmes (EAPs)** and other support options in providing help to individuals struggling with alcohol misuse. These programmes can be a lifeline for employees seeking help and can contribute hugely to the general wellbeing of the workforce. Finally, we will discuss the legal and ethical considerations surrounding workplace alcohol use, including alcohol testing policies, confidentiality and compliance with relevant laws and regulations.

As we embark on this journey to transform our workplaces, let us remember that the goal is not only to reduce the prevalence of alcohol-related harm but also to create environments where employees thrive, organisations prosper and wellbeing is valued.

NOTES

1 World Health Organization, 2022 (www.who.int/news-room/fact-sheets/detail/alcohol).
2 National Institute on Alcohol Abuse and Alcoholism, 2023 (www.niaaa.nih.gov/news-events/research-update/alcohol-related-deaths-which-increased-during-first-year-covid-19-pandemic-continued-rise-2021#:~:text=The%20figure%20above%20shows%20the,(from%2099%2C017%20to%20108%2C791)).
3 Scottish Government, 2023 (www.gov.scot/news/slight-rise-in-alcohol-specific-deaths/#:~:text=The%20figures%2C%20published%20by%20National,male%20deaths%20unchanged%20at%20836).
4 Adfam, 2020 (https://adfam.org.uk/files/one-in-three.pdf).
5 Home Office, 2012 (https://assets.publishing.service.gov.uk/media/5a7b71c1e5274a319e77f56d/ia-minimum-unit-pricing.pdf).

1 The Hidden Epidemic *Understanding Alcohol Problems in the Workplace*

1.1 THE PREVALENCE OF WORKPLACE ALCOHOL USE

During the past fifteen years of supporting employers to resolve alcohol-related challenges, I've observed a major shift in both mindset and approach. Ten years ago, workplace discussions about alcohol were more likely to centre around social drinking and bragging rights, rather than alcohol's effect on employee health and wellbeing. When companies did address alcohol misuse, the focus was typically on meeting safety and compliance requirements. While 90% of employers acknowledge that alcohol is an issue within their organisation, many have failed to treat it as a wellbeing concern. However, in the aftermath of the pandemic, more employers are beginning to recognise the importance of promoting a healthier relationship with alcohol as part of their broader wellbeing initiatives.[1]

According to the National Institute on Alcohol Abuse and Alcoholism (NIAAA), approximately 15 million Americans have an **alcohol use disorder (AUD)**, and most of them are employed. Here in the UK, over 10 million of us regularly drink at above the recommended maximum guidelines. Excessive drinking represents a considerable risk to health; worldwide, it is responsible for over 5% of all deaths.[2] These statistics alone should raise a red flag for HR managers. We cannot afford to underestimate the impact of alcohol misuse in our workplaces.

To grasp the magnitude of the issue, let's first acknowledge the prevalence of drinking among the workforce. It's a phenomenon that cuts across industries, from bustling corporate offices to factory floors, healthcare facilities to construction sites. It's not confined to any one sector, nor is it limited to any particular job title (although some occupations demonstrate higher risks than others). The reach of this epidemic is expansive. Contrary to the stereotype of the dishevelled, unemployed 'alcoholic', most people who drink heavily or binge (76% and 79%, respectively) are in full- or part-time employment. As alcohol is the highest-risk factor for death, disease and ill-health among the working-age population, it makes sense to establish just who is at risk and why.

Research links job-related stress with alcohol consumption and industries with high-stress environments, such as finance and healthcare, often report higher rates of alcohol misuse. The pressure to perform in these sectors, coupled with long working hours, can drive employees to seek solace in alcohol. Conversely, industries like information technology and academia may have lower rates of alcohol misuse, but the consequences can still be detrimental. Even in workplaces where alcohol use is not overt, it can lurk beneath the surface, affecting productivity, morale and

DOI: 10.1201/9781032648927-2

employee wellbeing. In a healthcare survey, 15% of employees admitted to being drunk at work[3] and around one in every twenty UK workers attend work with a hangover at least weekly.

We need to understand how alcohol use is tied to our culture and traditions. For example, in some European countries, it's common for colleagues to socialise after work over a glass of wine or beer. This can be a harmless way to build relationships and unwind, but it can also escalate into problematic drinking when not properly managed. Additionally, alcohol is often given as a reward for achievement or to celebrate success. Given alcohol's status as a risk factor for illness and death, and its major contribution to social care and crime costs,[4] this is surprising. Could you imagine, for example, offering a carton of cigarettes as a Christmas gift to clients? Or taking the team out for a few lines of cocaine to celebrate winning a big contract?

For those who don't drink, this alco-centric environment can have a huge impact on their wellbeing and self-confidence. For employees in recovery from an AUD, their sobriety may be at risk if their work environment encourages or even demands participation in alcohol-centred events.

According to the British Medical Association,[5] employees are more likely to drink frequently and to excess than the unemployed, and people in managerial and professional occupations drink to a greater extent than manual workers. Excessive drinking by employees is associated with negative workplace consequences, such as lowered productivity and increased presenteeism, costing the UK economy over £7.3 billion annually.[6] Additionally, over 17 million sick days are taken in the UK each year as a result of excessive drinking. A 2016 evidence review[7] into the effectiveness of alcohol control policies associated job-related stress with alcohol consumption and stated that 3.5% of employees report having to work extra hours to cover for the alcohol-related absence of colleagues.

Let's be clear, we're not talking about alcohol dependence, these figures largely do not relate to the small group of people who drink who are psychologically or physically addicted, although this minority is represented in the statistics; 80% of workplace alcohol problems are related to people who are not dependent on alcohol. That's right: social drinkers, 'normal' drinkers, all of us who drink are at risk of the consequences of excessive drinking. Here's a question for you to personally reflect on:

QUESTION FOR REFLECTION

Has *your* drinking ever impacted your work?

Consider hangovers, absenteeism, under performance, regretted behaviours etc. from throughout your work history. If you don't drink, has someone else's drinking affected you at work in some way?

It's not really possible to over play the impact of excessive drinking on our society. Alcohol misuse is a causal factor in over 200 medical conditions. It is also strongly linked to mental health problems including anxiety and mood disorders. This is a bi-directional relationship, with alcohol used as a coping mechanism in response to negative feelings and consumption leading to a depletion in key neurotransmitters, increasing feelings of anxiety and low mood. Psychiatric disorders such as schizophrenia have also been linked to excessive drinking. Another domain of alcohol harm is the impact of drinking on third parties through drink-driving, violence, impaired parenting and the burden on families supporting a loved one with alcohol dependence.

To put things into perspective: pre-pandemic *more working years were lost to alcohol than the ten leading causes of cancer death combined.*[8]

1.1.1 The Post-Pandemic Workplace

In March 2020, the World Health Organization (WHO) announced a global pandemic due to the spread of the COVID-19 virus. In the UK, this resulted in a national lockdown, commencing on 23rd March 2020. At that time, whilst all non-essential retail outlets and licensed premises were required to close, off-licences remained open as alcoholic drinks were listed as essential products, in opposition to WHO advice to restrict alcohol availability. The year 2020 saw an increase in off-trade alcohol sales of 24.4%, with those who bought the highest quantities *prior* to lockdown increasing their purchases the most.

In 2022, there were over 10,000 deaths from alcohol-specific causes registered in the UK. This represents a 32.8% increase on the data from 2019 (7,565 deaths).[9] These figures only include deaths that are *a wholly attributable direct consequence of alcohol consumption*, such as alcohol-related liver disease.

In the same year, there were a further 20,900 *alcohol-related* deaths—where alcohol was a key component of the health condition leading to the death—and almost 1 million alcohol-related hospital admissions.

The year 2022 also saw around 220 drink-drive deaths (that's 17% of all road deaths) and 6,700 injuries from drink-drive incidents.

These increases are perhaps unsurprising since researchers have found that the presence of stressors such as isolation, grief and economic downturns lead to increased drinking and potentially, the development of AUDs. Alcohol is used as a coping mechanism for negative feelings, and drinking contributes to feelings of anxiety and low mood.[10]

The unprecedented switch to home working allowed alcohol use during the working day to proliferate. With no daily commute, alcohol fulfilled a need to mark the transition between the working day and home life. Travel restrictions also reduced concerns about staying sober in order to drive. Home drinking provided an instant boost for an increasingly anxious population and the internet was bombarded with alcohol-positive memes extolling the virtues of alcohol as a coping strategy or an antidote to work pressures.

1.1.2 The Rise of Remote Work

Against this backdrop, much has been written about changes in alcohol consumption since the start of the pandemic. There is a consensus among researchers that between one in five and one in three of those who drank alcohol before the pandemic *increased* their consumption, and around one-third *reduced* their drinking during the first lockdown. What's interesting is that it was people *who were already drinking heavily* (i.e. regularly in excess of the recommended maximum and experiencing negative consequences as a result of their consumption) who started to drink more, whilst those who generally drink within the guidance maximum *started to drink even less*. So, whilst the figures seem to indicate an overall reduction in consumption, it's not an even spread across the population and has led to a rise in people seeking support for their increased drinking, post-pandemic.

The sudden onset of the pandemic forced organisations worldwide to adapt swiftly to remote work arrangements. What was initially considered a temporary response to a crisis has since become a long-term trend. Many employees now enjoy the flexibility of working from home, eliminating the need for daily commutes and providing a more comfortable work environment. It's crucial for managers to examine how this seismic shift in work arrangements has influenced the relationship between alcohol and work.

Virtual happy hours have become a staple of remote work, offering a way for employees to unwind and connect with colleagues from the comfort of their homes. However, the lack of in-person oversight makes it challenging to monitor alcohol consumption and behaviour.

Home working has blurred the boundaries between professional and personal life, and while working from the spare room offers numerous benefits, it also presents unique challenges related to isolation and stress. Loneliness and the absence of social connections that the workplace provides can take a toll on employees' mental wellbeing. As employees find themselves in their home environments, some may be tempted to incorporate alcohol into their workday. The ease of access to alcohol, reduced supervision and diminished social norms around drinking in remote work settings can contribute to an increase in alcohol consumption.

Commentators recognise the importance of establishing how drinking patterns have changed and for whom, and the likely impact on the workforce. Shortly after the first lockdown commenced, researchers began to explore the topic of health behaviour change, and alcohol consumption was a priority, given what was known about the potential impact of social isolation, job insecurity and national health concerns on drinking behaviours. Since the start of the pandemic, alcohol treatment agencies have identified a change in their client base. By the end of 2020, alcohol recovery services were already seeing an increase in the numbers of clients presenting for support in controlling their drinking but then came a shift in client demographics—more young people, more executives and particularly, more women.

1.1.3 Gender Disparities

Gender differences in alcohol use are worth noting. Historically, alcohol misuse has been more prevalent among men, but the gender gap is narrowing. Women are increasingly affected by AUDs, and this change has implications for workplace dynamics.

The reasons behind this shift are complicated and include changing societal norms, increased stress on women in the workforce and targeted alcohol marketing. We need to be attuned to these gender-related trends and adapt strategies accordingly to spot the negative consequences. The rise of the 'ladette' culture in 1990s' Britain encouraged women to compete with men in drinking bouts and uninhibited behaviour. Celebrity role models were pictured falling out of taxis at the end of a night of bingeing or knocking back beers with their peers and this style of drinking soon became the norm. Make no mistake though, this wasn't a random turn of events. Alcohol producers, concerned about falling revenues due to the buoyant rave scene, engineered the acceptance of 'drinking like the boys' and a generation of women was taught to feel empowered by alcohol.

Not only is alcohol advertising more targeted towards women, depicting camaraderie and the sisterhood, but also it hints at economic freedom and independence as facets of women's drinking. There has also been a massive increase in products designed to appeal to the female gaze. The popularity of events such as bottomless brunches and hen parties completes the picture of a new female drinking model. However, stigma and prejudices around female drinking remain, leaving women in a position of having positive expectancies around alcohol use whilst experiencing greater condemnation from society than their male counterparts if they mess up.

As this generation matures, they're also faced with research that shows that as women approach perimenopause and menopause, the risks that alcohol poses to their health and wellbeing are amplified. Alcohol is a leading cause of breast cancer in women;[11] it can increase the frequency and severity of menopause symptoms and can harm both the mother and the foetus during pregnancy.[12]

According to the WHO,[13] there are still considerably more men than women drinking at risky levels and subsequently suffering adverse health consequences, but women are catching up, with alcohol being responsible for 6.9% of the global burden of disease for men and 2% for women.

Women have traditionally been difficult to engage in alcohol recovery, they often experience many additional barriers including increased stigma, caring responsibilities and abusive relationships. The BMA[14] concluded that female high earners are particularly susceptible to developing alcohol problems and this is echoed by the rise in 'quit lit'—autobiographical accounts from high-flying women about their struggles with alcohol whilst trying to meet impossible standards of excelling in the workplace and in the home.

Finally, there is a troubling lack of research into the impact of alcohol on people who are trans. We know that alcohol impacts male and female bodies differently (due to size, hormones, processing times, etc.), but we know very little about how those differences impact someone who has transitioned. Researchers, take note. What we do know is that LGBTQ people are at greater risk of developing an **AUD** than the general population and that this increase is likely to be linked to victimisation and trauma.[15]

1.2 RISK FACTORS IN THE WORKPLACE

Not only does alcohol affect the wellbeing of individual employees, but it also poses risks to the organisations they work for. Before we delve into the potential negative consequences, let's examine some of the key factors that contribute to alcohol risk for the workforce.

1.2.1 Demographics

Different age groups and generations have varying attitudes towards alcohol and its role in their lives. Gen X remained the heaviest drinking age group as they passed through their 20s, 30s and 40s but now, in their 50s, the hard-drinking generation is embracing mindful drinking and up to a third describe themselves as 'sober curious'. This may in part be due to the influence of campaigns such as Dry January that support people who drink socially to try a brief spell of abstinence and reset their relationship with alcohol. The proliferation of 'grown-up' alcohol-free drinks has also provided alternatives that simply weren't available at the height of pub culture in the 1980s and 90s.

Younger employees may have different attitudes to drinking, with some prioritising work-life balance more than previous generations and around one in five under 25s claiming abstinence.[16] However, the influence of social media, where excessive drinking is often celebrated, can affect the alcohol-related choices of millennials and Gen Z employees. A report into young adult drinking[17] concluded that despite increased awareness of alcohol-related risk, many young people see binge drinking as an important facet of working life. A quarter of these cite stress relief as the reason for excessive consumption. Those most at risk are employees in manual jobs and young professionals, whilst public sector employees are less likely to partake in binge drinking. Interestingly, despite this, it is still the over 40s who are most likely to drink every day.

Cultural norms and socioeconomic factors within different racial and ethnic groups can impact alcohol use and importantly, increase the stigma around alcohol use and misuse. This can make it difficult for some individuals to acknowledge problems and seek help. These barriers can mean that people in some minority groups don't access services until they have experienced significant alcohol-related harm. Whilst some groups are less likely to use alcohol problematically, others are exposed to higher risk. Irish and Sikh men, for example, experience higher rates of alcohol-related liver disease than other ethnicities.[18] Sadly, there is a lack of research on the intersection of ethnicity alcohol use and the workplace.

To complicate things further, there are huge disparities in alcohol-related harm across the UK. The lowest rates of alcohol-related mortality are mainly found in the south of England whilst the Northeast and Northwest experience twice as many alcohol-specific deaths. In Scotland, the statistics are even more stark: deaths are six times higher in the most deprived areas compared to the most affluent.[19] This doesn't mean that people in deprived areas are drinking more, in fact, they drink less than those in the least deprived quintiles, but they experience more harm. Several factors influence this, including quality of housing, nutrition and healthcare. There are differences too between UK countries, as shown in Table 1.1.

TABLE 1.1
Alcohol Specific Death Rates and Alcohol-Related Hospital Admissions by UK Country

Country	Alcohol-specific death rate*	Alcohol-related hospital admissions
England	15.0	976,425
Northern Ireland	18.5	10,000
Scotland	22.6	35,781
Wales	17.7	54,900

*Figures for 2022 per 100,000 population[20], Alcohol Change UK

1.2.2 The Nature of the Work

The type of work itself can be a contributor to alcohol misuse. High-stress jobs, such as those in emergency services, healthcare or finance, often come with long hours, intense pressure and demanding expectations. Employees in these roles may turn to alcohol as a coping mechanism to manage the stress and anxiety that accompany their daily responsibilities. For example, consider the life of a healthcare worker in a busy urban hospital. Long shifts, the constant influx of patients and the emotional toll of dealing with life-and-death situations can create an environment ripe for alcohol misuse. The pressure to perform at the highest level at all times can be overwhelming, and some healthcare professionals may turn to alcohol as a means of escape. Indeed, in the short term, drinking alcohol will flood the brain with feel-good chemicals within just a few minutes of the first sip. It's a muscle relaxant and a disinhibitory, seemingly the perfect antidote to a stressful day. Research also shows that those who work for more than forty-eight hours a week are more likely to drink at higher levels than those who work forty hours or less. In fact, long working hours are associated with an 11% increased likelihood of increased drinking.[21]

Similarly, in the world of finance, traders and investment bankers face tremendous pressure to make split-second decisions that can have millions of pounds (or dollars) at stake. The intense competition and relentless drive for success can lead some to use alcohol as a way to unwind and cope with the constant demands of the job. What's more, these coping strategies are often encouraged. Alcohol is used in other ways in a corporate environment too, with drinking being a feature of corporate risk-taking. Where alcohol is sanctioned and even encouraged by corporate leaders, the impacts are amplified.[22]

In military service too, there is a strong association between alcohol use and 'decompression'—the practice of mandating a time-out of several days for personnel returning from active duty. Alcohol is an integral part of these events and there is evidence that it increases the normalisation of excessive drinking as a coping strategy.[23]

Research conducted in the UK[24] examined seventy-seven occupations and identified the industries where employees were most likely to experience heavy alcohol

TABLE 1.2
High- and Low-Risk Occupations for Alcohol-Related Mortality (UK)[25]

	Highest alcohol-related mortality by occupation		Lowest alcohol-related mortality by occupation
1	Bar staff, publicans and managers of licensed premises, waiters	1	Gardeners and groundspeople
2	Seafarers	2	Counter clerks, clerical officers
3	Butchers, meat workers	3	Managers (Office, retail, wholesale, transport, IT etc.)
4	Labourers in building and woodworking	4	Childminders, nursery nurses, primary education professionals, educational assistants
5	Musicians, entertainers	5	Metal workers and allied trades
6	Floorers and tilers, bricklayers, masons, painters & decorators	6	Prison service, security guards & related occupations
7	Chefs/cooks	7	Medical practitioners
8	Civil servants	8	Clergy
9	NCOs and other ranks	9	Farmers
10	Sales representatives	10	Vocational and industrial trainers

consumption and risk of dependence. Perhaps unsurprisingly, publicans, licensees, bar staff and others in the hospitality trade topped this list, whilst the clergy, scientists (including physicists, geologists, biochemists and medical practitioners) and primary/nursery educators proved to have the lowest risk.

Table 1.2 states the ten most and least likely industries correlated to alcohol-related mortality. It's important to understand these and other risk factors to implement preventative actions, as well as initiatives to reduce existing harm. Of course, not all heavy consumption translates into mortality. It's important that we also consider the wider impact on productivity, absenteeism and employee wellbeing.

1.2.3 Workplace Practices

Equally important are the workplace practices that shape employees' attitudes towards alcohol. Organisational norms around socialising, team bonding and celebrating achievements can inadvertently encourage excessive alcohol consumption. The infamous 'happy hour' culture prevalent in many workplaces can normalise drinking as a way to relax or connect with colleagues. Company-sponsored events and conferences often feature alcohol as a central component, creating an environment where excessive drinking can go unnoticed.

Alcohol is also often used to entertain clients and attract new business. This expectation or requirement to drink alcohol has been touted as a perk of the job but for some, it can exacerbate alcohol problems, whilst for those who would rather not drink, it is exclusionary and can be a bar to advancement or worse, derail their sobriety. Many firms take things further, offering a drinks trolley or having an onsite bar as an employment benefit, but they may be missing the mark. Around 20% of under 25s don't drink at all,[26] and wellness packages, private healthcare and flexible working are all rated far more highly by employees than free booze.

But even where alcohol is not a regular feature of the work environment, for most companies in the UK, the office Christmas party is a different matter. Up to 85% of employers provide alcohol for these events, either directly by offering free drinks as part of the party package or informally by purchasing drinks for colleagues during the event. These practices are fraught with risk and can leave the organisation vulnerable to litigation. Even where alcohol isn't provided, colleagues' behaviour in an alcohol-fuelled environment can lead to problems. Fortunately, however, in most cases, these celebrations pass uneventfully and are an expected and accepted component of work life.

CASE STUDY

An altercation broke out at a workplace Christmas party between two employees, who were both dating another employee. A wine glass was thrown by one, cutting the face of the other and one of the protagonists was dismissed, whilst the other received a final written warning. The dismissed employee took the employer to a tribunal for unfair dismissal and her complaint was upheld.

(*Westlake v ZSL London Zoo* ET/2201118/2015)

In recent years, many of these events have moved away from focusing solely on alcohol-based festivities and embraced a wider range of activities or at least the array of alcohol-free options available has improved. A survey by the Chartered Management Institute[27] found that 42% of employees surveyed (particularly those aged sixteen to thirty-four) felt that work parties should be organised around activities that don't involve alcohol.

There is clear evidence that where permissive drinking practices are coupled with a lack of clear policies and education around alcohol use, organisations are likely to see an increase in alcohol-related issues. When employees are unsure about the boundaries and expectations regarding alcohol, it can lead to risky behaviours and an increased likelihood of problems.

1.2.4 The Working Environment

Jobs that involve shift work, irregular hours or frequent travel can disrupt the normal routines and social connections that act as protective factors against alcohol misuse. The sense of isolation that can result from these work patterns may lead some employees to seek solace in alcohol. Work stressors coupled with fatigue have been shown to correlate with increased alcohol consumption, particularly where the person believes that alcohol will relieve either the stress or fatigue.[28] Over a quarter of people who drink claim that workplace stress makes them drink more.[29] Factors such as job insecurity, low autonomy, sick building syndrome and excessive working hours all contribute.

The following checklist identifies key work environment risk factors. Consider which of these apply to your organisation. To maintain a healthy and productive workforce, we also need to consider the range of potential impacts.

The after-effects of alcohol use include tiredness and low mood, partly due to poor sleep but also due to its effects on brain activity, which can be felt the following day. This can be exacerbated if the nature of the work is heavy and physical or if employees work long hours.

1.3 AREAS OF IMPACT

1.3.1 Health and Wellbeing

It is well-established that alcohol misuse can lead to dozens of physical and mental health issues, including liver disease, cancer, heart problems, depression and anxiety. These health concerns not only diminish the quality of life for affected individuals but also result in increased absenteeism and decreased productivity. Studies show that the consequences of excessive consumption include concentration and attention difficulties, reduced productivity, tiredness and mistakes. Still, social drinking among employees is a common and generally positive aspect of work life. Teambuilding, camaraderie and corporate hospitality are all positive effects of drinking culture. However, for those who don't drink or who have trouble controlling their consumption, these practices can be either risky to their sobriety or exclude them completely. It's important therefore, to examine the role of alcohol in your workplace and assess where it acts as a positive factor and where it may cause difficulties.

TABLE 1.3
List of Workplace Risk Factors for Excessive Drinking

Shift work
Travel away from home
Working remotely
Long working hours
Physically dangerous conditions (at height, underground, extremes of heat and cold, working with utilities)
Frontline with demanding or aggressive public
Low autonomy
Low job security, enforced change
'Permissive' culture around alcohol
Personal conflict at work
Frequent organisational change or sector uncertainty
Exposure to traumatic events
Exposure to personal safety risks (security, police, armed forces)
High turnover of staff
High responsibility with low autonomy

1.3.2 Safety Concerns

Alcohol impairs cognitive and motor skills, increasing the risk of accidents and injuries in the workplace. This is particularly concerning in industries where safety is paramount, such as construction, manufacturing and transportation. In these environments, the consequences of alcohol impairment can be catastrophic. Even a momentary lapse—forgetting a hard hat or a harness or neglecting to flick a safety switch could lead to injury, endangering not only the individual responsible but also their colleagues and bystanders. Statistics underscore the gravity of this issue: over half of fatal accidents at work involve falls from height[30] and alcohol-related injuries account for around one-tenth of the total impact of alcohol on health.[31]

1.3.3 Career Consequences

Of course, alcohol misuse can also have long-term consequences for an individual's career. Whether through poor job performance, absenteeism or a deteriorating work ethic, a worker may be faced with lead to disciplinary actions or even dismissal. Additionally, the stigma associated with alcohol problems can damage an employee's professional reputation and hinder their career advancement. Should an employee also receive a conviction for an alcohol-related offence, such as drink-driving or affray, this could have a reputational impact on both them and the company, as well as put their employment at risk.

1.3.4 Emotional and Social Impact

The impact of alcohol misuse extends beyond the workplace, affecting an individual's personal life and relationships. Family dynamics may suffer, and social connections can unravel as alcohol takes precedence over other priorities. The emotional toll on someone struggling with their alcohol use and their loved ones cannot be overestimated. If your employee is dealing with the alcohol misuse of a loved one, as around 5.2 million adults in the UK are, they are likely to be faced with a variety of challenges that can impact their day-to-day functioning, wellbeing, finances and more.

A third (34%) of those surveyed said they have been negatively affected by a friend or family member's drinking, drug use or gambling for more than ten years and a third (27%) between four and ten years.

Just over a third of people (31%) report that their loved one's drinking/drug taking or gambling negatively affects their working life, with 6% needing to take time off/reducing hours, 6% having to take sick leave more frequently and 22% struggling to concentrate/have a lower level of productivity when they're at work.

Just under half of those surveyed (45%) felt that they had been disadvantaged in some way because of their family member or friend's drinking, drug use or gambling. The most frequently cited disadvantages were: financial (24%), health (20%) and employment (14%).

(Adfam, 2021, Overlooked)

1.3.5 Decreased Productivity

Alcohol-related absenteeism accounts for around 5% of all absences. That's a nice, round measurable figure but it's not the whole picture. Much more difficult to quantify, but just as damaging to businesses, is presenteeism, where employees are physically present but mentally impaired due to alcohol use. Hangovers and even intoxication at work are an issue for many workplaces and 40% of employers agree that alcohol is a recognised cause of low productivity. This decrease in output can have a direct impact on an organisation's bottom line. Twenty-five per cent of employees say that drugs or alcohol have affected them at work, with the majority admitting to decreased productivity as a result.[32]

1.3.6 Increased Healthcare Costs

If your organisation provides healthcare for employees, whether via EAPs, counselling or insurance premiums, the health issues resulting from alcohol misuse among the workforce, may have led to increased healthcare costs including hospitalisations, rehabilitation programmes and mental health treatments. It would be easy to assume that this is a concern only for the small minority of workers who are identified as having an alcohol dependence, but these costs are largely incurred by employees whose drinking is classed as hazardous or high risk rather than dependent. In other words, those who we consider to be 'social drinkers', 'normal drinkers', adults who may rely on alcohol a little too much from time to time, to help with life's ups and downs.

1.3.7 Legal and Liability Risks

Alcohol-related incidents such as accidents or altercations can expose organisations to legal and liability risks. Failure to address this proactively may result in costly legal battles and damage to your organisation's reputation. Disciplinary action can be costly too. Tribunal costs for cases of unfair dismissal can be substantial so it's important that you have robust policies and practices in place to prevent and minimise litigation costs.

1.3.8 Damaged Workplace Culture

Employees may feel uncomfortable or unsafe in an environment, where there is a pro-alcohol culture, leading to reduced morale, higher turnover rates and difficulty attracting new staff. Where employees are required to take on extra duties or 'cover' for colleagues, resentment can build, harming an organisation's reputation and bottom-line.

1.4 THE ROLE OF LEADERSHIP

As leaders set the tone for what is acceptable within an organisation, their approach to alcohol can influence the entire workforce. Leaders who model responsible drinking can have a positive impact. When employees see their leaders drinking in moderation and prioritising responsible choices, it sends a clear message about expectations. Conversely, where the leadership team implements different rules for different employees or is lax about policy adherence or exhibits drinking practices that would be banned for staff, a very different, but just as clear message is sent.

1.4.1 Inclusivity

Leaders also play a vital role in ensuring that employees who choose not to drink or have a history of alcohol-related issues are not marginalised. Inclusive leaders create an environment where everyone feels comfortable and valued, regardless of their relationship with alcohol. Sober shaming, the practice of excluding or coercing peers who choose not to drink, has become a problem in some industries that have traditionally had a permissive attitude towards employee drinking. This can be a very divisive form of workplace bullying but is rarely brought to the attention of management. Once identified, it should be tackled promptly to avoid both workplace conflict and a potential legal battle.

1.4.2 Addressing Issues Proactively

Employees who are struggling with alcohol misuse (or are caring for a loved one who is in active addiction) are at their most vulnerable. A lack of compassionate support or a refusal to respond to difficulties caused by this can have a negative impact on the individual's path to recovery and their capacity to meet their work obligations. Your organisation's ethos will be evident in your response to such situations. If managers aren't trained to recognise signs of alcohol misuse, and importantly, don't have the confidence to tackle alcohol issues sensitively, problems can grow unnoticed and unchallenged.

1.4.3 Zero-Tolerance Policies

Zero-tolerance policies are common in safety-sensitive industries. These policies typically prohibit any alcohol consumption before or during work hours, with strict consequences for violations. This is driven by the need to ensure the safety of employees and the public and we will explore the evolution of zero-tolerance in a future chapter. Even a small amount of alcohol impairment can lead to disastrous accidents in work settings. However, there also needs to be some balance. Zero-tolerance measures are aimed at ensuring safety not as a means of punishing staff. Too often, unnecessarily punitive regulations restrict employee freedoms, leading to dissent, dissatisfaction and ultimately employees leaving the organisation. A recent survey[33] found that over 80% of workers in safety-critical roles support zero-tolerance policies, including drug and alcohol testing. Despite this, half of staff surveyed also believe that responsibility for health and safety lies with management, not with employees.

1.4.4 Permissive Alcohol Culture

In contrast, some workplaces have a much more permissive attitude towards alcohol use and a lifestyle that embraces socialising over drinks, with alcohol readily available in office spaces. In these environments, alcohol can be seen as a way to stimulate creativity, build team spirit and relieve stress. However, the risk of excessive drinking and its associated consequences remains a concern and research consistently shows that liberal workplace policies around alcohol consumption are associated with higher rates of employee drinking.[34]

NOTES

1 BACP, 2023. Alcohol at Work: A Heady Cocktail in BACP Workplace. *BACP Journal* July Issue, 8–13.
2 World Health Organization, 2019. Global Status Report on Alcohol and Health and Treatment of Substance Use Disorders.
3 Aviva, 2008, May. UK Employees Admit That Regular Drinking Affects Their Jobs (https://www.aviva.com/newsroom/news-releases/2008/05/uk-employees-admit-that-regular-drinking-affects-their-jobs-4048/).
4 In the UK, a total of £27.4 billion in 2022, including £4.91 billion to the NHS and healthcare, £14.58 billion to crime and disorder and £2.89 billion to social services; Institute of Alcohol Studies, 2024. The Costs of Alcohol to Society (https://www.ias.org.uk/report/the-costs-of-alcohol-to-society/).
5 British Medical Association, 2017. Alcohol, Drugs and the Workplace—The Role of Medical Professionals (https://www.bma.org.uk/media/1067/bma_alcohol-and-drugs-in-the-workplace-_oct_2019.pdf).
6 Institute of Alcohol Studies, 2024. The Costs of Alcohol to Society (https://www.ias.org.uk/wp-content/uploads/2023/03/The-costs-of-alcohol-to-society.pdf).
7 Public Health England, 2016. The Public Health Burden of Alcohol and the Effectiveness and Cost-Effectiveness of Alcohol Control Policies: An Evidence Review (https://www.gov.uk/government/publications/the-public-health-burden-of-alcohol-evidence-review).
8 Public Health England, 2016. Working Years of Life Lost Due to Alcohol (https://assets.publishing.service.gov.uk/government/uploads/system/uploads/attachment_data/file/574053/alcohol_public_health_burden_statistics.pdf).
9 Office for National Statistics. Alcohol-specific deaths in the UK: registered in 2022.
10 Koob, G. and Kreek, M. J., 2007. Stress, Dysregulation of Drug Reward Pathways, and the Transition to Drug Dependence. *American Journal of Psychiatry* Aug; 164(8), 1149–1159.
11 Cancer Research UK briefing (https://www.cancerresearchuk.org/about-cancer/causes-of-cancer/alcohol-and-cancer).
12 Foetal Alcohol Spectrum Disorder, 2022. National Institute for Health and Care Excellence [QS204] (https://www.nice.org.uk/guidance/qs204/resources/fetal-alcohol-spectrum-disorder-pdf-75547414426309).
13 Alcohol factsheet, World Health Organization (https://www.who.int/news-room/fact-sheets/detail/alcohol).
14 British Medical Association, 2017. Alcohol and the Workplace—The Role of Medical Professionals (https://www.bma.org.uk/media/1067/bma_alcohol-and-drugs-in-the-workplace-_oct_2019.pdf).
15 Hughes, T., et al., 2016. The Influence of Gender and Sexual Orientation on Alcohol Use and Alcohol-Related Problems. *Alcohol Research* (https://pmc.ncbi.nlm.nih.gov/articles/PMC4872607/).

16 Alcohol Change UK, 2024. Continued Decline in Drinking No Cause for Complacency (Blog) (https://alcoholchange.org.uk/blog/continued-decline-in-youth-drinking-no-cause-for-complacency).
17 Demos, 2016. Youth Drinking in Transition (https://demos.co.uk/research/youth-drinking-in-transition/).
18 Institute of Alcohol Studies, 2020. Ethnic Minorities and Alcohol (https://www.ias.org.uk/wp-content/uploads/2020/12/Ethnic-minorities-and-alcohol.pdf).
19 Public Health Scotland, 2017. Monitoring and Evaluating Scotland's Alcohol Strategy (https://publichealthscotland.scot/population-health/improving-scotlands-health/monitoring-and-evaluating-scotlands-alcohol-strategy-mesas/monitoring-report/).
20 Alcohol Change UK, 2025. Alcohol Statistics (https://alcoholchange.org.uk/alcohol-facts/fact-sheets/alcohol-statistics).
21 Public Health England, 2016. The Public Health Burden of Alcohol and the Effectiveness and Cost-Effectiveness of Alcohol Control Policies: An Evidence Review (https://assets.publishing.service.gov.uk/media/5b6c5703ed915d3119112af6/alcohol_public_health_burden_evidence_review_update_2018.pdf).
22 Guo, et al., 2024. Alcohol Culture and Corporate Risk-Taking. *Research in International Business and Finance* 67.
23 Osborne, A., et al., 2022. Military Service and Alcohol Use: A Systematic Narrative Review. *Occupational Medicine* Jun 8; 72(5), 313–323.
24 Thompson, A. and Pirmohamed, M., 2021. Associations Between Occupation and Heavy Alcohol Consumption in UK Adults Aged 40–69 Years: A Cross-Sectional Study Using the UK Biobank. *BMC Public Health* 21, Article number: 190 (2021).
25 Adapted from: Romeri, E., Baker, A. and Griffiths, C., 2007. Alcohol-Related Deaths by Occupation, England and Wales, 2001–05. *Health Statistics Quarterly* 35, 6–12 (www.statistics.gov.uk/statbase/product.asp?vlnk=6725).
26 Talbot, D., 2023. The Sober Myth: Are Young Adults Really a Generation of Non-Drinkers? Drinkaware (https://www.drinkaware.co.uk/research/research-and-evaluation-reports/the-sober-myth-are-young-adults-really-a-generation-of-non-drinkers).
27 Managers Pulse Point Poll, 2023. *Don't Make Booze the 'Main Event' at Work Christmas Parties After Survey Found a Third of Managers Witnessed Bad Behaviour*. Chartered Management Institute (https://www.managers.org.uk/about-cmi/media-centre/press-releases/dont-make-booze-the-main-event-at-work-christmas-parties-after-survey-found-a-third-of-managers-witnessed-bad-behaviour/).
28 Frone, M. R., 2016. Work Stress and Alcohol Use: Developing a Testing a Biphasic Self-Medication Model. *Author Manuscript* (https://pmc.ncbi.nlm.nih.gov/articles/PMC5228600/).
29 Alcohol Change UK, Alcohol in the Workplace. Factsheet (https://alcoholchange.org.uk/alcohol-facts/fact-sheets/alcohol-in-the-workplace).
30 British Safety Council, 2024. Falls Top the Cause of Workplace Deaths in 2023 (https://www.britsafe.org/safety-management/2023/falls-top-the-causes-of-workplace-deaths-in-2023).
31 Chikritzhs, T. and Livingston, M., 2021. Alcohol and the Risk of Injury. *Nutrients* Aug 13; 13(8), 2777.
32 Alcohol Change UK. Alcohol in the Workplace. *Factsheet* (https://alcoholchange.org.uk/alcohol-facts/fact-sheets/alcohol-in-the-workplace).
33 The Dräger UK Safety at Work Report, 2022 (https://www.draeger.com/Content/Documents/Content/Safety%20at%20Work%202022.Digital.pdf).
34 Thorrisen, et al., 2022. Are Workplace Factors Associated with Employee Alcohol Use? The WIRUS Cross-Sectional Study. *BMJ Open* (https://pubmed.ncbi.nlm.nih.gov/36229146/).

2 Cause for Concern

The Cost of Ignoring Alcohol Misuse

2.1 INTRODUCTION

In our ongoing exploration of alcohol misuse within the workplace, it becomes increasingly evident that the consequences reach far beyond individual employees. This chapter considers the myriad costs associated with alcohol misuse within organisational settings, taking into account the impact on individuals, families, communities and society at large. These encompass not only financial implications but also staff turnover, reputational issues, employee health and safety concerns and the profound impact on work performance and productivity. But first, we'll provide a brief historical perspective on alcohol and work and consider how drinking practices have influenced attitudes and drinking behaviours.

2.1.1 Historical Context

The human relationship with alcohol spans millennia, tracing back to ancient civilisations. As a species, we have a complicated relationship with alcohol, and it contributes both positively and negatively to society. It has acted as a social lubricant, fostering connections among individuals, strengthening bonds and facilitating trade; it has played a key role in rituals and gatherings, enhancing social interactions and has been a symbol of joy and celebration. Various societies have employed alcohol for its perceived medicinal properties; it has been prescribed to treat a variety of ailments and is believed to possess therapeutic effects.

Alcohol has also played a central role in religious ceremonies and celebrations. But while drinking has brought people together, it has often been accompanied by excessive consumption, leading to social disruption, health problems and accidents. Excessive drinking has a deep-rooted history of leading to adverse social and health consequences, a trend that continues today.

The historical context of alcohol use underscores its dual nature—a substance capable of enhancing human experiences but, when misused, leading to detrimental outcomes. To understand the contemporary landscape of alcohol use in the workplace, it's essential to explore its historical roots. Throughout history, alcohol has played an important and sometimes contradictory role in the workplace, reflecting broader societal attitudes and practices.

There is nothing unequivocal about it, and it's difficult to strike a balance between the positive aspects on one hand and the potential negatives on the other. As we start to explore this topic, here's a question for personal reflection:

DOI: 10.1201/9781032648927-3

QUESTION FOR REFLECTION

How would you describe your relationship with alcohol?

Has its contribution to your life been mostly positive or negative? If you don't drink, consider other ways in which you may have come into contact with alcohol and what the impact has been.

2.2 ANCIENT HISTORY

Studies of ancient pottery and art show us that alcohol has been produced since at least 9,000 BC, and one theory suggests that its production was responsible for the change in human civilisation from nomadic hunter-gathering to stable agrarian economies. In his book, Uncorking the Past,[1] historian Patrick McGovern hypothesises that from China to the Americas, across Africa and in Europe, cereals were probably first farmed for the production of alcoholic beverages rather than directly for food.

Originally made from honey, wild yeast and fruits, alcohol production methods became more varied and inventive as early humans developed a taste for this mind-altering substance. At the height of the ancient Egyptian civilisation, alcohol was a staple, depicted in art and early writing, and beer was a common drink for adults and children alike. Throughout Egypt's cultural dominance, manual workers were paid in beer, beginning a long association between alcohol and the workplace. For the upper classes, this was supplemented with wines. The Egyptians even had three different Gods associated with alcohol, and with the rise of the Greek and Roman empires, Dionysus (Bacchus to the Romans) was hailed as the God of wine and pleasure. In fact, all over the world, early civilisations assigned deities to protect and represent their relationship with alcohol.

2.3 PRE-INDUSTRIAL ERA

From early medieval times to the industrial revolution, alcohol had a central place in work and daily life. In agricultural communities, for example, workers might consume alcohol during harvest festivals or communal gatherings as a form of celebration and relaxation, but in feudal times, they were also often paid in beer by the local lord. Alcohol was seen as a social lubricant that fostered camaraderie among labourers and ale was even drunk in Christian churches right up until the 14th century.

Thereafter the alehouse grew in popularity as a place to buy and drink beer and became central to village life in Britain. Larger towns and cities housed taverns (for wine drinking) and inns (for the well-off). Despite this, ales and other beverages were watered down and it's likely that the average Briton of 500 years ago drank less than the current chief medical officer's recommended maximum of fourteen units a week.

Contrary to popular myth, most water supplies were perfectly safe to drink in the Middle Ages and beyond. Rather than being consumed for intoxication or for the lack of a viable alternative, alcohol was a ready source of calories and therefore energy. Estate owners would provide alcohol to serfs and indentured labourers to ensure they had energy for a day's work. However, the addition of hops to create a more stable and much stronger 'beer' from around 1400 onwards also leads to an increase in drunkenness so this positive aspect of alcohol use was tempered by its potential for excess. In some cases, heavy drinking among workers could lead to decreased productivity and workplace accidents.

By the end of the 15th century, whisky was in production in Scotland and after the dissolution of the monasteries by King Henry VIII, the practice of distillation formerly conducted by monks,[2] became more widespread providing much stronger alternatives to ale or beer. Despite this, alehouses continued to proliferate, firstly because of the potential for profit, secondly, because rising unemployment in the 16th century led to workers travelling greater distances in search of work, needing both somewhere to stay and a means of networking.

With the invention of the printing press and the expansion of writing in the 15th century, methods of production could be communicated more widely and trade via the Silk Route consolidated alcohol as an international commodity.

2.4 RUM RATIONS AND GIN LAWS

Picture the scene: it's 1650, and you're a sailor on a long voyage. What are you going to drink? There's no fresh water to hand, so your thoughtful captain provides a daily gallon (that's four litres) of beer for each crew member. Regularly transporting such large quantities of liquid across the seas was prohibitive, so in 1655, half a pint of rum was substituted and quickly became the drink of choice for able seamen.

Intoxication below decks was an increasing problem and by 1740, the rum was watered down. By the mid-19th century, the ration had halved to a quarter pint, in a further attempt to reduce intoxication. This ration, known as 'grog' (from whence the term 'groggy'), continued to be served daily to sailors in the Royal Navy until 1970.

Back on dry land, in 18th-century London, a change in the law made gin cheaper than beer. Falling food prices and a growing trade class meant that the population had more disposable income to spend on trendy spirits from the continent.

At its peak in the 1720s, it was estimated that gin consumption in London amounted to around a pint a week for every adult and child in the city.[3] Within a short time, imported gin was blamed for all society's ills, including poverty, idleness and debauchery, leading to more legislation being rushed through parliament in an attempt to stem the tide of cheap gin production. Hogarth's iconic prints of Gin Lane and Beer Street contrast the two beverages, depicting the decay of London due to the foreign import alongside thriving industry and commerce under the influence of beer.

2.5 INDUSTRIALISATION AND THE TEMPERANCE MOVEMENT

The industrial revolution brought huge changes to the workplace, including longer hours, harsh working conditions and a growing urban workforce. In this context, alcohol use began to take on a more negative connotation as excessive drinking could interfere with productivity and safety. Employers had to balance the benefits of fostering social bonds with the need to maintain a productive workforce. That said, by this time, beer drinking was firmly embedded as a working-class pastime with public houses acting as social clubs, places of entertainment and trade and the only community space in which to socialise away from home or work.

The temperance movement, which gained momentum in the 19th century, sought to curb alcohol consumption in society. Proponents argued that alcohol was a destructive force that led to inefficiency, absenteeism and slacking in the workplace, and moral decay at home. Many employers supported the temperance movement's efforts to reduce alcohol use among workers.

2.6 PROHIBITION AND ITS AFTERMATH

Although many UK towns were already dry on Sundays, the start of the First World War in 1914 saw the first national restrictions on the sale of alcohol across the UK with public house opening hours restricted to a couple of hours in the afternoon, for 'lunch' and a few hours in the evening, closing at 10.30 pm. These restrictions, welcomed by the temperance movement, continued after the war.

Across the Atlantic, the temperance movement welcomed national prohibition in 1920 in the United States, and this was to last for thirteen years, although laws banning or curtailing the sale of alcohol had been in place in various states for the previous half century. During this period, the production, sale and transportation of alcoholic beverages were banned. Whilst a victory for prohibitionists, consider the economic effect of prohibition: the closing of breweries, distilleries and saloons; loss of advertising, transportation, bottling and barrel-making revenue and the impact on other allied trades was a huge blow to the economy.

However, it was a progressive and radical step, and generally a popular one. Its aims were laudable: to eliminate the negative social and economic consequences associated with alcohol, particularly for women and children, who, although not allowed in saloons themselves were often the victims of drunken behaviour and poverty bought about by excessive consumption by their menfolk.

Whilst this policy led to a massive decrease in alcohol consumption in the short term, it also gave rise to underground speakeasies and illegal alcohol production. Ultimately, as a result of increased crime, corruption and the proliferation of illicit stills and breweries prohibition was considered a failure. Throughout this period in the workplace, some employers continued to provide alcohol to employees as a way to maintain morale and loyalty. Here in the UK, the temperance movement continued to exert its influence.

2.7 POST-PROHIBITION ERA

After the repeal of prohibition in 1933, alcohol once again became legally available, and its role in the workplace continued to evolve. Employers began to adopt policies to regulate alcohol use on the job, particularly in safety-sensitive industries. This was backed by legislation further consolidating on and off-licencing opening hours, in response to public health concerns about the illness of 'alcoholism' and the effects of intoxication. By the time the 1967 Breathalyser Act came into force, banning driving under the influence of alcohol, it was widely accepted that alcohol could profoundly affect physical and mental health in both the short and long term. In 1974, the Health and Safety at Work Act[4] was implemented and placed a legal duty on employers to control workplace drinking in safety-critical jobs.

Within a couple of decades, workplace **EAPs** emerged, identifying alcohol dependence as a treatable medical condition. Employers and unions collaborated to provide assistance to employees struggling with AUDs, marking a shift from punitive measures to supportive interventions.

2.8 CONTEMPORARY WORKPLACE ALCOHOL CULTURE

Today, workplace alcohol culture varies widely across industries and organisations. Some workplaces maintain a strict zero-tolerance policy, while others may have more permissive attitudes towards alcohol use. Understanding this diversity is essential for HR and OH managers seeking to address alcohol-related issues.

As you will have seen from Chapter 1, the global COVID-19 pandemic has raised new challenges for workplaces with respect to alcohol-related harm and it remains to be seen what the longer-term implications will be for both employers and employees.

2.8.1 The Financial Impact of Workplace Alcohol Misuse

One of the most immediate and tangible consequences of alcohol misuse is its financial toll. Although we may at first think of lost productivity and absenteeism—and rightly so: here in the UK alcohol costs the economy over £7 billion annually[5]—the costs associated with alcohol misuse are broad and affect organisations in numerous and often subtle ways.

2.8.2 Absenteeism and Presenteeism

Absenteeism, characterised by employees frequently being absent due to alcohol-related issues, places a clear financial burden on organisations. Such absences result in lost productivity, and the costs of hiring temporary replacements often fall on employers. A report by the Institute of Alcohol Studies found that over 17 million days of sickness absence are taken each year in the UK due to excessive drinking.[6] This accounts for around 4–5% of all absenteeism and loss of a colossal 167,000 working years annually.

Several studies show this direct correlation between alcohol consumption and absence, with young employees and males at greatest risk.[7] Surprisingly, there is no correlation between the total *amount* of alcohol consumed and the likelihood of work absence, rather it is heavy episodic drinking, commonly known as bingeing, that accounts for much of this absenteeism. Young adults are more likely to indulge in heavy episodic drinking, than their older colleagues, who statistically favour daily but more moderate drinking. This accounts for the increased rates of absenteeism among young workers. Young women are less affected by this phenomenon, with twice as many young men as young women taking time off work for alcohol-related reasons.[8]

In a recent survey, the work environment was cited by 27% of survey respondents as the reason for excessive drinking.[9] This was attributed either to a permissive drinking culture in the workplace or the stresses of the modern workplace; excessive workload, high staff turnover, low autonomy and poor management all being cited as circumstances that can drive people to drink.

Consequences of alcohol-related absenteeism include increased workload for other employees, increased overtime costs and a decline in productivity and employee satisfaction. While the financial impact on organisations may be significant, the personal toll on the employees struggling with absenteeism is equally concerning. When you consider that alcohol is responsible for over sixty medical conditions and diseases and has been identified as the leading risk factor for premature mortality, ill-health and disability among 15–49-year-olds in the UK,[10] it's easy to see how important a role alcohol can play in the cost of absenteeism.

Presenteeism, on the other hand, involves employees physically present at work but physically or psychologically impaired due to alcohol use, through working whilst either intoxicated or hungover. Alcohol-related productivity losses are challenging to quantify precisely, but they can be substantive. A report by the Institute of Alcohol Studies calculated the cost of workplace hangovers at over £1 billion annually. Employees who misuse alcohol may experience reduced cognitive function, decreased attention to detail and slower task completion. Inefficiencies in workflow and reduced output can cause errors and missed deadlines and can negatively impact an organisation's bottom line. Elevated insurance premiums and the implementation of additional workplace safety measures, such as safety training and equipment upgrades, can also contribute to financial burdens.

The long-term consequences of persistent alcohol misuse can affect an employee's livelihood. Employees who struggle to control their drinking may experience a lack of career advancement, resulting in reduced earning potential and job satisfaction. Alcohol-related issues can also impede an employee's ability to participate in professional development opportunities, hindering their growth within the organisation.

A 2019 study[11] found that 42% of employees had attended work with a hangover or whilst under the influence and on average they rated their own workplace performance to be 39% lower than usual.

Over a third (36%) had suspected one or more colleagues of being hungover or under the influence in the previous six months. Respondents cited the impacts of this to include being personally less productive, less effective teamwork and lower morale, increased stress and longer working hours.

Interestingly, the prevalence of hangovers/intoxication at work increases with earnings, with those earning over £45,000 p.a. at greatest risk.

For some, hangovers at work will be a rare occurrence, but over time, for those struggling to control their consumption, this can become a regular feature of their working life. Research suggests that levels of presenteeism vary by industry, with the hospitality sector being particularly susceptible to negative consequences. Understanding the extent and impact of absenteeism and presenteeism can help organisations target policies and workplace practices to reduce the risks.

QUESTIONS FOR REFLECTION

In your workplace, which do you think takes a greater financial toll: absenteeism or presenteeism and why?

Consider how presenteeism may affect safety or business critical roles, the impact on insurance, liability and legal costs against the cost of hiring in temporary labour, offering overtime or dealing with reduced productivity and efficiency from absenteeism.

How could you determine which of these is a bigger issue in your workplace and what it's costing the business?

You may want to collect data on absenteeism and productivity before implementing an alcohol programme, so that you have baseline data for comparison later.

2.8.3 Legal and Liability Costs

Another area of concern is the cost of legal battles and liability claims resulting from alcohol-related incidents in the workplace, such as accidents or injuries. Organisations may be held accountable for incidents involving employees under the influence of alcohol, leading to litigation, compensation payouts and reputational damage. However, the financial costs are only part of the story; injuries, trauma and other consequences for workers and their families can lead to long-lasting physical and emotional challenges.

With an average of around 45,000 drink-drive convictions annually in England and Wales, around 270 road deaths and over 6,000 casualties[12] attributable to excessive consumption, any organisation that employs drivers or whose staff commute needs to be aware of the potential risks. Chapters 6 and 7 explore best practice approaches to alcohol-related incidents and legal liability in more depth. A Norwegian study concluded that between 4% and 13% of employees experienced alcohol-related sexual harassment in work settings. Workplaces with more employer-sanctioned drinking events and tolerance for intoxication at work-related drinking occasions were correlated with increased incidents of harassment.[13]

2.8.4 Healthcare Expenditures

Employees struggling with AUDs may require rehabilitation and wellbeing services, including mental health support or counselling. If caught early, short-term interventions are often sufficient to support lasting change but where an established problem has developed, longer-term support may be appropriate. These healthcare costs, which can escalate rapidly, may fall under the employers' health insurance plans or covered by an EAP. The financial burden of treatment can strain an organisation's healthcare budget and the personal toll on employees seeking treatment is considerable. However, the majority of workplace alcohol problems involve employees *who are not alcohol dependent.* What's more, an estimated one in four workplace accidents are related to alcohol use.[14]

2.8.5 Staff Turnover and Morale

A huge 35% of people surveyed say that they've noticed colleagues under the influence of alcohol or other substances at work[15] and a 2020 study found that one in six employees experienced harm as a result of their coworkers' drinking.[16] Naturally, employees may experience strained relationships with colleagues or supervisors as a result, which can disrupt teamwork and hinder collaboration. A study by the Institute of Alcohol Studies also found that about one in four employees had attended work with a hangover and one in ten had done so in the last six months.[17] This may lead coworkers to feel frustrated, overburdened or demoralised when they have to pick up the slack for colleagues struggling with alcohol-related issues. Low morale can lead to decreased engagement, which, in turn, affects productivity and job satisfaction.

Employees struggling with alcohol misuse are more likely to leave their jobs voluntarily or have their employment terminated due to performance or conduct issues. Additional recruitment and training costs for new employees to replace them are then incurred, not to mention the loss of knowledge and experience.

CASE STUDY

A warehouse operative whose tasks included driving a forklift truck received a conviction for drink-driving resulting in a twenty-two-month driving ban. As a driving licence is not a prerequisite for driving a forklift, his employer was happy to keep him in employment. However, on renewal of the company's insurance, the insurer explained that they could not provide cover for the employee because of his conviction. The employer was forced to terminate the operative's employment as he could not carry out his duties without company insurance.

Organisations with a reputation for a permissive attitude to drinking may struggle to attract skilled staff. The 'work hard, play hard' trope has lost its shine, and prospective employees may be discouraged by an unhealthy workplace environment and the perception that the organisation does not value their wellbeing. Alternatively, certainly, in sectors such as hospitality, employees may be attracted to a work environment precisely because of the company's tolerant approach to drinking.

The good news is that there is a growing number of campaigns and initiatives that can support people to make changes to their drinking. A flourishing 'sober curious' movement has led to increased acceptance of mindful drinking and sobriety as positive lifestyle choices. For example, research into the effectiveness of the UK's annual Dry January challenge[18] shows that participants report a range of wellbeing benefits from just thirty-one days of abstinence.

These include better mental wellbeing and physical health, weight loss, improved sleep quality and memory, improved concentration and energy levels and feeling more in control of their alcohol consumption. Most participants self-identify as social drinkers, in fact, anyone identifying that they may be dependent on alcohol is advised not to take part in the campaign but to seek medical advice and support if they wish to change their drinking. It's clear that even moderate drinking can impact a range of health outcomes.

2.9 EMPLOYEE HEALTH AND SAFETY

2.9.1 Accidents and Injuries

Alcohol problems in the workplace pose risks to health and safety, contributing to a large number of work-related accidents.[19] Employees under the influence of alcohol are also more prone to workplace injuries, putting themselves and their colleagues at risk and an alarming 15% of employees admit to intoxication at work.[20] These incidents can result in physical harm, loss of productivity and potential legal liabilities for the organisation. Whilst the construction and transport industries have traditionally borne the brunt of these risks, there is considerable evidence that alcohol-related injuries in the workplace are more widespread than we may think. Alcohol impairs cognitive function and decision-making so employees struggling with alcohol misuse may make poor judgements that can lead to unsafe practices, workplace hazards and errors.

CASE STUDY

In 2023, the Law Society Gazette[21] reported that an employment tribunal judge had received a formal reprimand for drinking alcohol during the working day. Changes in the judge's behaviour led colleagues to believe that she may have been intoxicated. Evidence indicated that the judge was indeed under the influence and the scheduled tribunal hearing could not be completed that day.

2.9.2 Safety-Sensitive Roles

In safety-sensitive and safety-critical industries, such as transportation and manufacturing, the risks associated with alcohol misuse are particularly pronounced. Even minor impairment can have huge repercussions, making strict alcohol policies and regulations imperative. In some industries, such as aviation, this is backed by legislation prohibiting or reducing alcohol use for several hours before commencing work.

There is also a case for employees in business-critical roles, such as those with financial responsibility, to be tested or face drinking restrictions in the workplace, especially as figures show that people in managerial and professional roles tend to drink more than blue-collar workers.

THREE KEY EVENTS THAT CHANGED WORKPLACE ALCOHOL LEGISLATION

Eltham Well Hall Rail Disaster, 1972

At the time of the derailment of a ten-coach diesel train from Margate to Kentish town, its driver had a blood alcohol concentration of more than three times the drink-drive limit. Five passengers and the driver were killed and over a hundred passengers injured. The subsequent Department of the Environment[22] report concluded that the driver:

> *'had grossly impaired his ability to drive safely by drinking a considerable quantity of alcohol both before and after booking on duty, including some shortly before leaving Margate, and some more in his cab during the journey'.*

The report into the tragedy recommended 'for cause' alcohol testing to be carried out on drivers and seconds in the event of any collision. Later, zero-tolerance and random testing regimes were instituted for railway workers.

Exxon Valdez Oil Spill, 1989

When the supertanker Exxon Valdez ran aground off Alaska, causing the largest oil spill in history, its captain was serving a driver's licence suspension after a conviction for driving under the influence of alcohol in 1988. The Exxon report into the accident found that Hazelwood, the captain, was under the influence of alcohol, although this was not considered to be a factor in the accident. Recommendations from the National Transportation Safety Board in the wake of this disaster included:

> *Implement an alcohol/drug program for seagoing employees that prevents such personnel from returning to sea until their alcohol/drug dependency problem is under control.*
>
> (Class 11, Priority Action) (M-90–30)

> *Train persons who monitor the alcohol/drug rehabilitation program in the recognition of recidivism after treatment, in the utilization of appropriate professional referrals and in the interpersonal skills necessary for competent rehabilitation supervision.*
>
> (Class 11, Priority Action)

The Murder of Lt Cdr Ian Molyneux, 2011

An able seaman fired six shots from an SA80 rifle in the control room of a Royal Navy submarine, killing his commanding officer and injuring three other officers. The inquest into the death of Lt Molyneux found that the sailor responsible had blood alcohol levels in excess of the drink-drive limit at the time of the incident.

In 2011, sailors were allowed to consume up to ten units of alcohol in the twenty-four hours before going on duty. This was subsequently reduced to five units, with no alcohol whatsoever in the ten hours preceding duty.

2.9.3 Litigation

There are several instances of employment tribunals concerning employee dismissal as a result of conduct or capability issues caused by alcohol consumption. Litigation is both costly and distressing and high-profile cases have highlighted the importance of establishing robust procedures and adequate risk management strategies to prevent and respond to alcohol-related incidents. Besides the financial costs of investigations, legal representation and potentially hiring and training additional staff to replace employees whose positions have been terminated, the reputational damage to the organisation can be harmful and the impact on employees involved can be life-changing.

CASE STUDY

An employee brought a successful claim against her employer after reporting rape by a manager after a Christmas party. There had been an expectation that employees would attend the event, with expenses in part paid for by the employer and a glass of prosecco was provided to employees on arrival.

During the event, the site manager became intoxicated and, on several occasions, sexually harassed and assaulted a member of staff (not the complainant). When challenged by another manager, he became abusive. The site manager was then seen talking to another, junior female employee, who appeared intoxicated. Despite the site manager's actions throughout the evening, his increasing intoxication and the intoxication of the female employee, no action was taken to protect her and no-one informed her of the site manager's actions towards other members of staff. This fact and her levels of intoxication meant that she was not in a position to *'perceive, let alone assess, any risks to her personal safety'*. Rather, *'she placed her trust in the Site Manager as a known senior colleague'*, according to the tribunal judge, *'and allowed him to accompany her in a taxi to her hotel'*.

The claim of harassment against the employer was upheld as it was deemed that the employer failed to protect the claimant and that the Site Manager's conduct was in the course of his employment.

(*P v Crest Nicholson Operations Limited,*
Employment Tribunal case number 3311744/2020)

2.9.4 Physical and Mental Health

For anyone struggling with alcohol misuse, the personal toll includes deteriorating physical and mental health. This can manifest as liver disease, cancers, cardiovascular problems and a reduced quality of life. Mental health can also suffer from anxiety, social anxiety and depression all being consequences of long-term excessive drinking. The correlation between alcohol problems and common mental health problems is well-established, with those diagnosed being twice as likely to report an AUD.

According to the Local Alcohol Profiles for England (LAPE)[23]

In 2017–2019, there were approximately 60,000 cancer registrations related to alcohol consumption

In 2020, there were over 432,000 years of life lost due to alcohol-related conditions

In 2022, there were 7,912 alcohol-specific deaths (wholly attributable to alcohol) in England, an increase of 4.7% from the previous year.

2.9.5 Relationships

Beyond the workplace, the consequences of alcohol misuse ripple outwards to affect all aspects of society. Alcohol misuse often strains personal relationships, leading to conflicts, separations and divorce. Affected individuals and their loved ones may experience isolation and social alienation. Divorce rates are higher for marriages where one or both partners have an AUD.[24] Research from Adfam, the national charity supporting anyone affected by a loved one's addiction, suggests that *one in three adults* in the UK is affected by a loved one's drinking,[25] with up to 4% of families also experiencing abuse from loved ones who drink.[26] Studies show that this can severely impact wellbeing—psychologically, physically, socially and financially. Family members are often reluctant to access help for themselves, focusing on changing the drinker. They endure emotional distress, worry and anxiety, witnessing their loved one's decline. Drinking can also deplete family finances through excessive spending on alcohol, legal fees and reduced capacity to work. Supporting a loved one through alcohol addiction will also take a toll on an employee's wellbeing in the workplace. As a responsible employer, demonstrating an understanding of the profound impact that a loved one's drinking can have on family life and the health of your staff will provide an important lifeline for employees.

After the drink-driving conviction of Ant McPartlin, one half of the celebrity duo Ant and Dec, Declan Donnelly stated in an interview for *The Guardian* magazine that the duo nearly parted ways over the incident, almost ending a thirty-year collaboration. Donnelly described his partner's actions as 'indefensible'. McPartlin was imposed a £86,000 fine and banned from driving for twenty months for the offence in which he drove into two other vehicles.[27]

2.9.6 Impact on Children

Children in families affected by parental drinking may face neglect, abuse and disrupted home environments. The consequences can manifest in various ways, including academic struggles and emotional trauma. The charity NACOA[28] found that there are around 3 million children in the UK who have at least one caregiver who is experiencing an alcohol problem.

These children are more likely than the general population to witness domestic violence, develop an eating disorder, consider suicide, experience difficulties at school, develop an addiction or be in trouble with the police. Parental alcohol problems are also associated with poorer physical and psychological health.[29] Even children living with a parent who uses alcohol excessively but not dependently, of whom there are up to 2.5 million in the UK,[30] are at increased risk of short-term psychological harm. A 2017 report[31] highlighted the impact of non-dependent drinking on children, with young respondents citing more arguments, less parental attention, disrupted bedtimes and worry or embarrassment as a result of their parent's drinking. Given these startling statistics, it's important to acknowledge that you may have employees who were brought up in households where alcohol was used problematically and may still be experiencing the consequences of this.

2.9.7 Burden on Statutory Services

Let's not overlook the impact of alcohol on the NHS, which is often discussed, particularly with reference to alcohol-related incidents that require emergency medical attention. Emergency services, already stretched, must allocate resources to address these issues, and an added burden to the National Health Service is the almost 1 million hospital admissions for alcohol-related causes recorded in the UK annually. However, primary care physicians are also impacted by alcohol problems, from the responsibility of managing patients with alcohol misuse issues to accessing support for their loved ones, leading to annual healthcare costs in England alone of almost £5 billion.[32] Whether as a result of alcohol-fuelled accidents requiring urgent treatment or the longer-term treatment of chronic conditions caused by excessive drinking, every facet of our healthcare system is burdened.

But as with so many aspects of alcohol culture, it's more complicated than that. The criminal justice system also bears the brunt of alcohol-related issues, including arrests, court proceedings, probation and incarceration. Additionally, almost half of all violent crime in the UK is alcohol-related, contributing to a staggering annual financial burden of over £14 billion.[33] These are not just issues relating to the night-time economy—alcohol is also implicated in domestic abuse and sexual assaults. Let's be clear: alcohol does not cause domestic violence or sexual assault, it is always a choice made by a perpetrator, but an alarming number of victims state that their assailant was under the influence of alcohol at the time of the offence. Alcohol doesn't make anyone do something they're not capable of whilst sober, it just reduces their regard for the consequences.

NOTES

1 McGovern, P., 2009. *Uncorking the Past: The Quest for Wine, Beer, and Other Alcoholic Beverages.* University of California Press.
2 The brewing process involves adding yeast to a base, such as a grain or fruit to produce fermentation. Distillation—the manufacture of spirits—is an additional process in which alcohol is separated from the fermented beverage, creating a more potent drink.
3 Nicholls, J., 2009. *The Politics of Alcohol.* Manchester University Press.
4 See Chapter 7 for further information about the Health and Safety at Work Act.
5 Institute of Alcohol Studies factsheet, 2024. Economy (https://www.ias.org.uk/factsheet/economy/).
6 Institute of Alcohol Studies Briefing, 2020. Alcohol in the Workplace (https://www.ias.org.uk/wp-content/uploads/2020/12/Alcohol-in-the-workplace.pdf).
7 Public Health England, 2016. The Public Health Burden of Alcohol and the Effectiveness and Cost-Effectiveness of Alcohol Control Policies: An Evidence Review (https://www.gov.uk/government/publications/the-public-health-burden-of-alcohol-evidence-review).
8 Schou, et al., 2014. Alcohol-Related Sickness Absence Among Young Employees: Gender Differences and the Prevention Paradox. *European Journal of Public Health* 24(3).
9 Alcohol Change UK. Alcohol in the Workplace Factsheet (https://alcoholchange.org.uk/alcohol-facts/fact-sheets/alcohol-in-the-workplace).
10 Office for Health Improvement and Disparities, 2024. Alcohol Profile (https://www.gov.uk/government/statistics/alcohol-profiles-for-england-february-2024-update).
11 Institute of Alcohol Studies, 2019. Financial Headache. The Cost of Workplace Hangovers and Intoxication to the UK Economy (https://www.ias.org.uk/wp-content/uploads/2020/06/rp35062019.pdf).
12 Gov.UK, 2022. Reported Road Casualties in Great Britain, Final Estimates Involving Illegal Alcohol Levels: 2020 (https://www.gov.uk/government/statistics/reported-road-casualties-in-great-britain-involving-illegal-alcohol-levels-2022/reported-road-casualties-in-great-britain-involving-illegal-alcohol-levels-2022).
13 Lund et al., 2023. Alcohol-Related Sexual Harassment in the Workplace: A Between and Within Industry Analysis. *Drug and Alcohol Review* Nov; 42(7), 1764–1772.
14 Institute of Alcohol Studies Briefing, 2020. Alcohol in the Workplace (https://www.ias.org.uk/wp-content/uploads/2020/12/Alcohol-in-the-workplace.pdf).
15 Considerate Constructors Scheme. Spotlight on . . . drugs and alcohol (https://ccsbest-practice.org.uk/spotlight-on/drugs-and-alcohol/#the_campaign).
16 Moan, I. and Halkjelsvik, T., 2020, Work-Related Alcohol Use and Harm to Others. *Substance Use & Misuse* 55 (https://www.tandfonline.com/doi/full/10.1080/10826084.2020.1801744).
17 Institute of Alcohol Studies Briefing, 2020. Alcohol in the Workplace (https://www.ias.org.uk/wp-content/uploads/2020/12/Alcohol-in-the-workplace.pdf).
18 De Visser, R., 2019. *Evaluation of Dry January 2019.* School of Psychology, University of Sussex (https://www.drugsandalcohol.ie/32647/1/R-de-Visser-Dry-January-evaluation-2019.pdf).
19 Institute of Alcohol Studies Briefing, 2020. Alcohol in the Workplace (https://www.ias.org.uk/wp-content/uploads/2020/12/Alcohol-in-the-workplace.pdf).
20 Institute of Alcohol Studies, 2020. Alcohol in the Workplace (https://www.ias.org.uk/wp-content/uploads/2020/12/Alcohol-in-the-workplace.pdf).
21 Hyde, J., 2023. Judge Warned Over Drinking During a Hearing Break. *The Law Society Gazette* (https://www.lawgazette.co.uk/news/judge-warned-over-drinking-during-a-hearing-break/5114918.article).
22 Report on the Derailment that occurred on 11th June 1972 at Eltham (Well Hall) Station in the southern region British Railways; London: HMSO, 1973.

23 Gov.uk, 2022. Local Alcohol Profiles for England: Short Statistical Commentary (https://www.gov.uk/government/statistics/alcohol-profiles-for-england-february-2024-update/alcohol-profiles-for-england-short-statistical-commentary-february-2024).
24 Cranford, J. A., 2014. DSM-IV Alcohol Dependence and Marital Dissolution: Evidence from the National Epidemiologic Survey on Alcohol and Related Conditions. *Journal for the Study of Alcohol and Drugs* (https://pmc.ncbi.nlm.nih.gov/articles/PMC4002864/).
25 Adfam, 2021. Overlooked (https://adfam.org.uk/wp-content/uploads/2024/06/Overlooked.pdf).
26 Adfam, 2012. Between a Rock and a Hard Place (https://adfam.org.uk/wp-content/uploads/2024/06/Between_a_rock_and_a_hard_place_-_Project_report.pdf).
27 Hattenstone, S., March 30, 2019. Interview in Guardian Weekend Magazine (https://www.theguardian.com/tv-and-radio/2019/mar/29/ant-and-dec-we-considered-splitting-up-over-indefensible-drink-driving).
28 National Association for the Children of Alcoholics.
29 NICE, 2011. NICE Guideline: Diagnosis, Assessment and Management of Harmful Drinking (High-Risk Drinking) and Alcohol Dependence.
30 All Party Parliamentary Group on Children of Alcoholics, 2017. A Manifesto for Change (https://www.issup.net/files/2018-04/Children-of-Alcoholics-A-Manifesto-for-Change-pages.pdf).
31 Institute of Alcohol Studies, 2017. Like Sugar for Adults (https://www.ias.org.uk/uploads/pdf/IAS%20reports/rp28102017.pdf).
32 Institute for Alcohol Studies, Economy fact sheet (https://www.ias.org.uk/factsheet/economy/).
33 Institute for Alcohol Studies, Economy fact sheet (https://www.ias.org.uk/factsheet/economy/).

3 All about Alcohol

3.1 INTRODUCTION

In the preceding chapters, we've set the scene and identified that alcohol can be detrimental to both employer and employee and it's now time to explore why. In this chapter, we will begin by explaining what alcohol is and how it affects the human body, defining some commonly used terms. We will explore the causes and risk factors associated with alcohol misuse, explore into the physical and psychological effects and outline the stages of addiction. Moreover, we will provide clear definitions of what constitutes an alcohol problem.

To underscore the impact of alcohol misuse on functionality, we'll explore how intoxication affects an individual's executive functions, response times, memory, thinking and communication abilities, all of which can have severe implications in a work setting.

We probably think that we know about alcohol, it's ubiquitous, it's all around us, we live in an alco-centric society. But there are many myths and misconceptions that can trip us up. This chapter will give you the confidence to challenge untrue assertions about alcohol and its potential for harm.

QUESTION FOR REFLECTION

What do you know about alcohol? How would you rate your knowledge?

After reading this chapter, consider what you've learned—has anything surprised you? Is there anything that you've learned that you're keen to share with others? If so, why?

3.2 WHAT IS ALCOHOL?

Alcohol is a generic term for a group of organic compounds containing one or more hydroxyl groups; (they have -OH in the formula) and this is responsible for the shared properties of alcohols. Their names end in 'ol', such as propanol, butanol, glycerol and methanol, but the alcohol that we drink is *ethanol*, with the molecular formula C_2H_5OH. Whether drinking vintage champagne or white cider, the *alcohol* in a drink is always the same. Of course, production methods, flavourings and strengths may vary but only one type of alcohol—ethanol—is sold legally in alcoholic beverages. Throughout this book when referring to ethanol, the generic term 'alcohol' will be used.

Most alcohols are not safe for human consumption but have a variety of commercial uses. The alcohols are colourless and (mostly) odourless. They are flammable but don't create smoke when they burn (making them useful fuels), water-soluble (making them helpful as solvents) and volatile (so they can disperse the fragrance of perfumes and then evaporate quickly). Also used as hand sanitisers, antiseptics and a whole host of other industrial applications, alcohols are a commodity used for much more than just drinking.

DOI: 10.1201/9781032648927-4

Ethanol is made via a process called fermentation. This involves the conversion of glucose (sugar) within grains or other plant material and yeast into ethanol and carbon dioxide. The fermentation process produces liquid that is up to about 15% ethanol. This is the process used in the manufacture of beers. A further process, distillation, can be utilised to evaporate and condense the fermented drink into a much more potent beverage, as can be found in spirits.

3.2.1 Alcohol Strength

Ethanol strength is measured in %ABV (alcohol by volume), in other words, the percentage of the volume of liquid that is pure ethanol. So, if a beverage contains 20% ABV then 20% or one-fifth of the drink is pure ethanol.

In the UK, alcohol is measured in SAU—**standard alcohol unit**—with one unit equivalent to 8 g or 10 ml of ethanol. The units in any alcoholic drink can be calculated using two variables—the volume in ml (how much liquid in millilitres) and the %ABV (what percentage of the liquid is ethanol). Every container of alcohol sold legally in the UK, whether a can, a bottle or a box, will display this information.

The volume and strength totals are multiplied together, and the result divided by 1,000 to give the unit content:

Calculating Units

Vol (ml) × %ABV/1000 = units
Example: a large (250ml) glass of 13%ABV wine
250 × 13 = 3250
3250/1000 = 3.25 units

Current UK guidance, published by the Chief Medical Officer in 2016,[1] gives three recommendations:

1. *A weekly guideline on regular drinking*
    - You are safest not to drink regularly more than fourteen units per week.
    - If you do drink this much, it's best to spread your drinking over three or more days.
    - The risk of developing a range of illnesses (including some cancers) increases with any amount you regularly drink.
2. *Advice on single episode drinking*
   To reduce your short-term health risks from a single drinking occasion:
    - Limit the total amount you drink on any single occasion.
    - Drink more slowly, with food and alternating with water.
    - Avoid risky places and activities.
    - Some groups are at higher risk and should be more careful of their drinking levels on any one occasion, including, young adults, the elderly, people with

low bodyweight or other health problems or who are using medicines or other drugs.

3. *A guideline on pregnancy and drinking*
   - It is safest not to drink at all if you are pregnant or planning a pregnancy.
   - Drinking during pregnancy can lead to long-term harm to the baby, the more you drink the greater the risk.

The guidance goes on to stress that there is no safe level of consumption of alcohol—the guidelines are for low-risk drinking.

3.2.2 The Nature of Alcohol

Alcohol is a *central nervous system depressant*—it slows down the way that the mind and body work. Normally consumed as a liquid, alcohol passes into the mouth, throat and stomach and small bowel where, unlike food, it is not digested but absorbed directly into the bloodstream. This is why blood and breath tests for alcohol work—if it had already been broken down by enzymes in the stomach and intestines, as food is, there would be nothing in the blood and breath to measure. Once in the stomach, a common effect of drinking is acid reflux. This is caused by alcohol's property as a muscle relaxant. The pyloric valve, the ring of muscle at the base of the oesophagus, relaxes, allowing stomach acid to slip through causing heartburn and reflux.[2]

Via the circulatory system, alcohol crosses the blood-brain barrier within five minutes. Once there, it can affect everything from our conscious actions such as movement and speech to our autonomic nervous system—our breathing, temperature control and heart rate. Finally, after about half an hour and a further trip round the body, it reaches the liver. This is where it is broken down by enzymes, first into a highly carcinogenic compound called acetaldehyde and finally into carbon dioxide and water. The liver can process, on average, about one unit per hour. When more than this is consumed, unprocessed alcohol remains in the blood, travelling around the organs and through all bodily systems until it can be broken down.

The impact of all levels of alcohol consumption on any individual will be determined by several factors, from age and gender to hydration levels and physical health. Some will be able to consume much more alcohol and over a longer period than others seemingly without any lasting effects. Similarly, some people will experience short-term effects such as hangovers and blackouts at fairly low levels of consumption, whilst others appear immune. In other words, alcohol affects everyone differently and there's no way of determining in advance who will be susceptible to which aspects of alcohol harm, if at all. Make no mistake, though, alcohol is a toxic substance and even though its effects may not be visible on the outside, its impact on the body can be devastating.

During the Dry January challenge in 2018, staff at Alcohol Change UK, the charity that runs the campaign began to receive emails from participants commenting that they'd noticed that their resting heart rate had decreased and wondered whether this was a side effect of going alcohol-free for a few weeks. The answer is yes!

Fitbits were a popular Christmas gift in 2017 and suddenly people were able to track for themselves the difference that not drinking made to their heart rate, sleep quality and more.

Within the human body, alcohol content is measured in ***blood alcohol concentration (BAC)*** and the level of BAC or the speed at which BAC rises and falls will differ from individual to individual, as will susceptibility to the many physical harms that alcohol can cause. Alcohol can impact many of our bodies systems, including cardiovascular, gastrointestinal tract, brain and nerves. It is a carcinogen (cancer-causing), a neurotoxin (a nervous system poison) and a teratogen (causes damage to the embryo and foetus). In total, alcohol has a causal relationship with over sixty different diseases and contributes to the development and progression of over two hundred others.

3.3 PHYSICAL AND PSYCHOLOGICAL EFFECTS

3.3.1 Alcohol's Impact on Neurotransmitters and Brain Function

Neurotransmitters are chemical messengers that transmit signals in the brain and play an essential role in regulating various bodily functions, including mood, cognition and behaviour. Alcohol disrupts the delicate balance of neurotransmitters, leading to both immediate and long-term effects. Individual differences, susceptibilities and genetics, as well as lifestyle factors, will determine the impact of alcohol on the individual.

3.3.1.1 Immediate Effects

1. **Enhanced GABA activity** Alcohol enhances the neurotransmitter gamma-aminobutyric acid (GABA). This results in sedation, relaxation and decreased anxiety. However, excessive GABA activity can impair coordination and decision-making. When the effects of alcohol wear off, GABA transmission is reduced, and anxiety increases.
2. **Inhibition of glutamate** Alcohol inhibits the glutamate, contributing to the sedative effect. This inhibition slows down brain activity, affecting alertness and cognitive function. As blood alcohol levels fall, glutamate increases, leading to greater restlessness.
3. **Release of Dopamine** Alcohol consumption triggers the release of this hormone, the brain's 'feel-good' neurotransmitter. This surge in dopamine reinforces the pleasurable effects of alcohol, making it rewarding, and encouraging continued use.

Alcohol is an **agonist** for GABA—it **increases** transmission of this neurotransmitter.

GABA is **inhibitory** in the brain—it slows down brain function, inducing relaxation.

Alcohol is an **antagonist** for glutamate—it **reduces** the levels of glutamate.

Glutamate is an **excitatory** neurotransmitter—it **increases** activity in the brain.

Alcohol is an **agonist** for dopamine, which can be either excitatory or inhibitory.

3.3.2 Long-Term Effects

1. **Neuroadaptation** Chronic alcohol misuse leads to neuroadaptation, where the brain adjusts its neurotransmitter activity to counter the effects of alcohol. Think of it as a see-saw effect. The brain compensates for the depressant effects of alcohol by decreasing inhibitory neurotransmission on the one hand and enhancing excitatory neurotransmission on the other to achieve equilibrium. Ultimately, this results in increased tolerance, upsetting the see-saw again and requiring more alcohol to achieve the same effects and bring the balance back to neutral.
2. **Diminished dopamine response** Over time, the brain's dopamine response to alcohol decreases. As a result, individuals may experience reduced pleasure from everyday activities, which can lead to depression and increased alcohol consumption to compensate.
3. **Impaired serotonin function** Alcohol misuse can disrupt serotonin, a neurotransmitter associated with mood regulation. This disruption can contribute to mood disorders, such as depression and anxiety.
4. **Alterations in glutamate** Chronic alcohol misuse can lead to glutamate rebound during periods of abstinence (withdrawal). Without the presence of alcohol, there is an excess of excitatory activity causing anxiety, irritability and cravings for alcohol. (That see-saw is out of balance again.) In severe withdrawal, these symptoms can escalate to seizures, delirium and organ failure.

Excessive drinking in the long term can lead to dependence – a set of symptoms characterised by a loss of control over consumption, despite evident harm, that is driven by complex biological and psychological factors. The neuroadaptive effects described earlier are only part of the story. They can lead to intense cravings for alcohol, which can become compulsive, leading individuals to prioritise alcohol over other aspects of their lives. Behaviour and mood will be disrupted in different ways depending on whether the person has not consumed alcohol for a few hours and may be approaching a withdrawal state, has 'topped up' and is not in need of an immediate drink, nor showing signs of intoxication or drinking to the point of inebriation. An employee who is alcohol-dependent is likely to exhibit fluctuations and inconsistencies in performance and demeanour often characterised by anxiety and inability to concentrate. The damage, both physical and psychological, is progressive and can have cumulative effects. In the early stages, individuals may experience mild consequences, such as hangovers or occasional lapses in judgement. However, as drinking continues it increases the risks and severity of harm increase.

Without intervention, late-stage alcohol misuse can result in severe physical and mental health issues, loss of employment, social isolation and diminished quality of life. The modal age for alcohol-specific deaths (i.e. wholly attributable to conditions caused by alcohol) is 55–59 years, with twice as many men as women making up the mortality rates. I reiterate: alcohol remains the biggest risk factor for death, ill health and disability among 15–49-year-olds in the UK.[3]

3.3.3 Alcohol Withdrawal

For the person who is physically dependent on alcohol, stopping drinking suddenly is neither safe nor advisable, as alcohol withdrawal symptoms can be dangerous and even fatal. Within a few hours, symptoms commence including sweating, hand tremors, nausea, headaches and anxiety. Within a couple of days, these may abate or may escalate to more serious consequences such as hallucinations and seizures. For those most seriously affected, delusions, known as *delirium tremens* can also start within two to three days.

These are characterised by confusion and hallucinations along with high blood pressure and a racing heart. Alcohol withdrawals can be life threatening but, under medical supervision a person can be detoxed from alcohol in seven to ten days. A word of caution, though, repeated withdrawal, even via controlled detox can also be dangerous, increasing the likelihood of complications. At this point, it's also helpful to clarify that removing alcohol from the body via detox is not a cure for dependence. It is simply a means of allowing the brain's neurotransmitters to regain equilibrium. Cravings and compulsions around alcohol use are much more difficult to overcome and require ongoing intervention and relapse prevention support. More on treatment options in Chapter 8.

3.3.4 Impact on Cells in the Brain and Body

Excessive drinking can damage neurons (that's brain cells to you and me), impairing cognitive function, memory and motor skills. Current estimates suggest that up to 70% of people with an AUD will experience some cognitive deficits, including difficulties with planning and organisation, problem-solving and critical thinking, and up to 40% may have alcohol-related brain damage.[4] Alcohol can also cause brain shrinkage. Yes, it really does kill brain cells. Long-term alcohol misuse can lead to brain atrophy, resulting in further cognitive deficits and increased vulnerability to neurological disorders.

3.3.5 Carcinogenic Effects

Alcohol is classified as a Group 1 carcinogen by the International Agency for Research on Cancer (IARC). It is known to cause several types of cancer, including cancers of the oral cavity, pharynx, larynx, oesophagus, colorectum, liver and breast.[5] Even moderate alcohol consumption has been linked to an increased risk of breast cancer in women, due to alcohol's impact on oestrogen levels. In 2020, an estimated 741,300 newly diagnosed cases of the above cancers were attributable to alcohol consumption.[6] This may not be the full picture, research is ongoing into whether alcohol is responsible for other cancers including those of the stomach, lungs, pancreas, kidney and urinary bladder.

> The absolute increase in cancer risk for one bottle of wine per week equals that of five (men) or ten (women) cigarettes per week.[7]

The mechanisms by which harm occurs are varied. Alcohol can damage cells and reduce the efficacy of cell repair; it increases levels of oestrogen and insulin, contributing to additional cell division, and it can increase access of other carcinogens (e.g. in cigarettes) to the mouth and throat.[8]

3.3.5.1 Liver and Metabolism

When considering alcohol's harm to the body, the liver is often the first organ we think of, and with good reason. The liver is responsible for metabolising this highly toxic substance and breaking it down into harmless carbon dioxide and water, ready for elimination. Chronic alcohol misuse can lead to liver diseases such as alcohol-related fatty liver, hepatitis, fibrosis and finally, cirrhosis. The liver is a remarkable organ, the second largest in the human body, with over 500 different functions. Its importance in supplying energy, maintaining body temperature, supporting the immune system, removing toxins and producing proteins (amongst other functions) cannot be overstated. So, when the liver is damaged by alcohol, other systems in the body, including the brain, are affected.

It's often said that the liver can regenerate and yes, it has a high capacity for repair and regeneration, but once damaged by alcohol, for example, when cirrhosis is present, the liver's ability to heal is compromised.

The liver is also a silent organ: it has no pain receptors, so symptoms of a damaged liver can be hard to spot or late in presentation, once considerable damage has occurred. Even with cirrhosis, which occurs usually after many years of excessive drinking, there may be sufficient liver function to support normal bodily activity. Often, it's only when liver failure sets in that symptoms become apparent. Despite this, diet and lifestyle changes can yield very positive results in reversing diseases such as fatty liver or alcohol-related hepatitis; cirrhosis cannot be reversed but immediate cessation of alcohol consumption can prevent further damage and improve life expectancy. If drinking continues, however, the five-year survival rate is below 50%.[9]

Liver function tests, blood tests to help determine the cause of symptoms or to monitor damage to the liver, will indicate levels of proteins and other substances that can determine liver health. However, it's important to understand that the accuracy of these tests can vary and that results within a normal range do not mean that no damage has occurred, only that, at the time the test was taken, the liver was functioning within normal parameters.[10] A more effective measure of damage is the FibroScan, which can determine the extent to which scarring (fibrosis) has occurred on the liver.

Alcohol affects metabolism, causing metabolic syndrome by altering insulin sensitivity and thereby contributing to weight gain. In other words, it slows down metabolism and reduces the body's ability to burn fat. It can also lead to the development of conditions like alcohol-related pancreatitis, hypertension and type 2 diabetes.[11]

3.3.6 Heart and Cardiovascular System

Alcohol also increases blood pressure, which over time can become chronic (hypertension), increasing the risk of heart disease, stroke and other cardiovascular problems. It can also weaken the heart muscles, leading to cardiomyopathy, a condition that impairs the heart's ability to pump blood effectively. Chronic drinking can damage the stomach lining, leading to gastritis and an increased risk of ulcers. It is also a common cause of acid reflux.

3.3.7 Cognitive Functioning and Alcohol-Related Brain Damage

Cognitive decline is a common feature of AUDs with current estimates suggesting that between 50% and 80% of dependent drinkers experience cognitive impairments.[12] Not only does it affect mood and memory but also executive functions such as attention, planning and organisation, task initiation and self-awareness. It can be difficult for someone whose drinking is excessive to fully realise its impact. Often, people whose drinking continues despite clear evidence that it is causing them harm are described as being 'in denial', but heavy drinking reduces awareness of difficulties and makes it harder for people to realise that they need help. If unchallenged, people with AUDs will continue to drink at harmful levels.

Alcohol-related brain damage risk starts at thirty-five units a week—that's about half a bottle wine or two pints of lager *per day*.

Drinking more than twenty-one units a week increases dementia risk.[13] That's just over two bottles of wine or about seven to eight pints of lager *per week*.

Employees struggling with alcohol-related issues may find it challenging to make informed decisions as alcohol impairs the ability to gather and process information necessary for effective decision-making. This impairment can lead to suboptimal choices in the workplace, such as poor resource allocation or missed opportunities.

Longer-term alcohol use can also impair problem-solving skills. Employees facing alcohol-related impairments often encounter cognitive hurdles when addressing intricate workplace issues, resulting in challenges such as delayed resolutions and ineffective strategies to plan and organise tasks. What's more, capability issues are progressive in nature.

3.4 MOTOR SKILLS

One study found that motor impairments—coordination, speed of movement and strength were—all impaired during the hours after blood alcohol levels returned to zero.[14] A systematic review also concluded that hangovers negatively affect attention, focus and working memory as well as other executive functions.[15] In other words, the hangover phase is associated with impairment to both physical and mental abilities.

In safety-sensitive roles, alcohol's impact on reaction times is of particular concern, potentially leading to increased safety risks and compromised emergency response.

3.5 MOOD AND BEHAVIOUR

Alcohol affects mood and behaviour in two phases, with its impact changing according to BAC. At lower levels, it typically acts as a stimulant, boosting sociability, confidence and mood—attributes that can be highly valued in certain job roles. However, as BAC rises, alcohol's depressive effects become more dominant, causing lethargy, confusion, mood swings and occasionally aggression.[16] This shift can understandably lead to more workplace conflicts and negatively affect morale.

3.5.1 Gender Differences

Gender is one of the largest factors influencing the impact of alcohol harm on the body. Men and women process alcohol differently.[17] Differences in metabolism, size body fat percentage and other factors will influence the impact on different genders. For example, when consuming the same quantity of alcohol, breath alcohol concentration rises to greater levels in women than in men. For every unit consumed, a male's breath alcohol concentration will increase by 7 µg whilst a female's will increase by 11.7 µg, often leading women to experience the effects of alcohol more intensely. This can lead to quicker impairment in decision-making and motor skills. Women may also be prone to adverse health effects at lower drinking levels and over a shorter period than men.

Women who drink heavily are at greater risk of alcohol-related brain damage at a younger age than men.[18] They may also experience higher levels of alcohol-related anxiety and depression and be more susceptible to liver problems, cardiovascular disease, blackouts and hangovers.[19] It's certainly true that as women approach perimenopause, their symptoms are likely to be exacerbated by drinking including fatigue, mood swings, poor sleep quality, weight gain, brain fog and anxiety. As one in three women say they drink more since perimenopause, it's important that they understand the potential negative impact that their drinking may have on their wellbeing.

For every 1,000 women who drink two or more units per day, an extra five will develop breast cancer over the next five years compared to the background risk.[20]

In contrast, men may be more likely to exhibit the loss of control associated with late-stage addiction. They are more likely to engage in heavy drinking and binge drinking and four times more likely to become alcohol dependent, which can lead to more severe health consequences and a higher likelihood of aggressive behaviour when intoxicated.

3.5.2 Age

The very young and the elderly are more susceptible to the negative effects of alcohol psychologically, physically and socially. The human brain is not fully developed until well into the third decade of life, exposing young people who drink heavily to behavioural and cognitive risks, though their bodies may be more robust and recovery times from drinking bouts shorter.

For young people who have a parent with an alcohol use disorder, this can be particularly problematic as they are more likely to develop an alcohol problem than their peers. What's more, issues with alcohol are likely to develop quickly in a phenomenon known as the telescoping effect.

At the other end of the scale, older employees may find that cognitive decline and increasing after-effects such as low mood and hangovers increasingly become features of their alcohol consumption. Over time, they may have developed a tolerance to alcohol and be drinking far in excess of the recommended maximum without realising.

TABLE 3.1
List of Potential Impacts of Excessive Alcohol Consumption in the Short and Long Term

Short-term	Short- and long-term	Long-term
Reduced motor control	Memory loss	Stroke
Nausea and vomiting	Disturbed sleep	Vitamin deficiency
Impaired coordination	Loss of muscle control	Impaired blood clotting
Intoxication	Increased vulnerability	Cancer of mouth, throat, breast, bowel
Interaction with medication	Reduced infection resistance	Ascites, pancreatitis
Disinhibition	Increased risk of miscarriage	Alcohol-related brain damage
Diarrhoea	Impaired sexual performance	Peripheral neuropathy
Risk of violence	Impaired sensation	Ulcers/inflammation of stomach
Risk taking behaviours	Impaired blood clotting	Poor self-care/hygiene
Disorientation, confusion	Hallucinations	Early onset dementia
Increased aggression	Lethargy and tiredness	Oesophageal varices
Acid reflux	Paranoia	Heart disease, hypertension
Blackouts	Low mood/mood swings	Dependence
Accidents	Depression	Liver damage
Hangover	Anxiety	Malnutrition
Euphoria	Increased accident risk	Foetal Alcohol Spectrum Disorder

3.6 WHAT IS AN ALCOHOL PROBLEM?

[Alcohol's impact on the brain] is more bewildering that the workings of the financial markets, economy or weather. It is a web of billions of exquisitely delicate, interconnected components not understood even in isolation. Alcohol's warping effects only multiply the difficulty.

Our brains are not all the same either, meaning our responses to alcohol can be dramatically different, differences added to by variations in our digestions of alcohol. We also respond differently at different times depending on our surroundings and state and past exposure. And our response also depends on our expectations, beliefs which vary between us and change.

(Cain, P., 2018, Alcohol Companion)

In order to address the issue effectively, we need to understand how alcohol problems develop. While there is no one-size-fits-all explanation, and no causal link has been identified, there are definite evidence-based correlations between alcohol problems and several common risk factors that may contribute to someone developing difficulties. But first, what is an alcohol problem?

There's a lot of confusion around alcohol terminology—alcoholic, dependence, **AUD** and binge—what does it all mean? Well, whilst some terms are used interchangeably, it would be helpful to standardise the use of terms in this book. According to guidance published by the **National Institute for Health and Care Excellence (NICE)**, harmful drinking is defined as

> *a pattern of alcohol consumption causing health problems directly related to alcohol. This could include psychological problems such as depression, alcohol-related accidents or physical illness such as acute pancreatitis. In the longer term, harmful drinkers may go on to develop high blood pressure, cirrhosis, heart disease and some types of cancer, such as mouth, liver, bowel or breast cancer.*

Alcohol dependence (see also **AUD**) is

> *characterised by craving, tolerance, a preoccupation with alcohol and continued drinking in spite of harmful consequences (for example, liver disease or depression caused by drinking).*

Dependence, AUD and harmful drinking are used in preference to the older, rather pejorative term 'alcoholic'.[21] Specific criteria for alcohol dependence can be found in the ICD-10[22] manual under the code F10. Collectively, for disorders characterised by excessive alcohol consumption, the term 'alcohol use disorders' is used. *Alcohol misuse* is used throughout this book to identify a pattern of drinking that is having a persistent negative effect on the life of the person drinking. AUD is used where an individual has received a diagnosis of an alcohol-related disorder.

Strong links have been identified between alcohol consumption and mental health problems including anxiety and mood disorders.[23] It can be hard to establish whether alcohol was the cause or consequence of poor mental health and in some cases is likely to be both. More profound psychiatric problems such as personality disorders and suicidal ideation have also been linked to excessive drinking.[24]

> Around 4% of the UK population is thought to be alcohol-dependent, although only about 18% of these are engaged with alcohol treatment services.

A helpful and simple (but not diagnostic) definition of an **AUD** is:

> *When alcohol is affecting your physical, psychological or social functioning.*

Note that this does not describe a pattern of drinking or an amount or a type or frequency of consumption. The issue is how alcohol affects the individual in their daily life. In other words, different people will be affected in different ways even at similar amounts of alcohol consumed. This is a helpful paradigm as it allows us to focus on the problems caused by alcohol, rather than the alcohol itself.

For example, if ten random adults each drank twenty-eight units of alcohol each week for ten years, it's possible that one of them would have a damaged liver, one may have experienced a relationship breakdown due to their drinking, one may struggle with anxiety and poor sleep, one may have a drink-drive conviction, one may have had an accident due to alcohol. A couple (or more) of our random sample may have noticed no ill effects, whilst others may see their lives unravelling. Alternatively, all might be well and thriving or all may be experiencing negative consequences. It's complicated. We do know that there is a dose correlation with alcohol problems—the more someone drinks, the greater the risks.

3.6.1 Stages of Addiction

Addiction is not a simple matter of in or out, it is rather a continuum with blurred boundaries, rather than a set of clearly defined stages. However, the progression towards addiction does follow a recognised path.

Many, many people initially consume alcohol socially without negative consequences. For example, lots of us occasionally enjoy a drink with friends after work without any apparent issues. Whilst not indicative that alcohol problems will develop, there are also many people who would describe themselves as social drinkers whose consumption patterns and levels suggest that they find it increasingly hard to control their drinking. This might include daily drinking or heavy episodic drinking patterns. Over time, individuals may build tolerance, requiring more alcohol to achieve the desired effect.

Regular alcohol consumption becomes a habit, often escalating in frequency to daily or almost daily use and a sense of unease if for some reason there is no alcohol available at the usual time or place. At this point, the person may begin to experience negative consequences such as missed work or strained relationships or exhibiting signs of stress.

Over time, the person may struggle to control their alcohol intake, often drinking excessively or at inappropriate times or places followed by unsuccessful attempts to cut down or stop.

Health problems may become more pronounced and are clearly attributable to alcohol use, although, for the person drinking, these consequences are often ignored or fobbed off as having other non-alcohol causes. This is a common feature of dependence. At this point, addiction can also lead to social consequences, including job loss, legal issues and social isolation. The workplace, then, offers an important opportunity to raise awareness of alcohol risk, educate employees and support improved wellbeing outcomes for the more than 10 million of us in the UK who drink at above low-risk levels.

3.7 MODELS OF ALCOHOL DEPENDENCE

There are several models that seek to determine why some people become dependent and others don't, often with similar patterns of drinking. Despite much research, we still don't know for certain why some people seem more susceptible to alcohol problems than others. The following are the current most prevalent models for addiction, but we may yet discover more or decide that all of these factors may play a role.

3.7.1 Spiritual Deficit

The idea of alcohol dependence as a form of demonic possession or spiritual void is rooted in cultural beliefs that have shifted over time. In ancient societies, behaviours linked to excessive drinking were often attributed to supernatural forces, with the belief that someone experiencing alcohol dependence was under the influence of malevolent spirits or demons. This interpretation offered a way to make sense of the suffering experienced by those who drank excessively, especially in the context of loss of self-control and behavioural changes framed as moral or social degradation.

By the mid-19th century, the temperance movement gained popularity, offering a lack of moral or spiritual guidance as the cause of addiction and promoting faith-based activities such as prayer and penitence as the cure and holding the individual accountable for their dependence.

As scientific knowledge about addiction and mental health advanced, both of these models have faded from mainstream thinking, although remnants of these belief scans still be found in some cultural or spiritual contexts. In some communities where religious beliefs strongly shape health views, individuals may still interpret alcohol dependence as a form of spiritual affliction, influencing how they seek help and support.

3.7.2 Disease Model

The disease model understands alcohol dependence as a chronic health condition rather than a personal weakness or moral issue. This approach suggests that genetic predisposition, biological factors and environmental influences all play a role in the development of an alcohol problem, leading some people to be more vulnerable to addiction. Similar to other chronic conditions like diabetes or heart disease, alcohol dependence is seen as a progressive disorder that impacts physical, psychological and social wellbeing over time.

Central to the disease model is the idea that alcohol dependence changes brain function, particularly in areas linked to reward and pleasure, resulting in compulsive drinking and a vastly reduced ability to control alcohol intake. Due to these brain changes, individuals may experience intense cravings and withdrawal symptoms, making it challenging to stop drinking without assistance.

This model has shaped modern treatment by emphasising medical and therapeutic support, including counselling, medications and peer support, rather than approaches based on judgement or punishment. By viewing alcohol dependence as medical condition, this model reduces stigma and encourages empathy, highlighting the need for holistic ongoing care.

3.7.3 Biopsychosocial Model

In the last forty or so years, this model has achieved prominence. It views alcohol dependence as a consequence of biological, psychological and social influences. Instead of pointing to a single cause, this model suggests that alcohol dependence arises from a combination of genetics (bio), mental health dynamics (psycho) and environmental (social) pressures, each contributing to an individual's unique experience with alcohol.

From a biological standpoint, genetics have a strong influence, as some individuals inherit a heightened susceptibility to addiction based on family history. Additionally, alcohol use impacts brain chemistry, particularly in areas related to reward and impulse control, which can lead to physical dependence. Over time, the brain may begin to rely on alcohol to stabilise mood, making it increasingly difficult for individuals to quit on their own.

Psychological influences also play a dynamic role. Emotional and mental health challenges—such as depression, anxiety, trauma or chronic stress—often drive individuals to use alcohol as a temporary relief or coping mechanism. Alcohol's effects on mood can create a repetitive pattern where drinking becomes the preferred response to negative emotions or stress.

Social and environmental factors shape drinking habits as well. Family and community norms can heavily influence a person's attitude towards alcohol; for example, growing up in an environment where drinking is accepted or encouraged can normalise alcohol use. Peer pressure may also encourage individuals to drink more frequently or heavily. Cultural and societal attitudes, as well as access to alcohol and economic factors, also contribute to patterns of alcohol dependence.

This model supports a holistic approach to treatment, recognising that a combination of medical, psychological and social interventions is often necessary to effectively address the diverse factors that underly dependence.

3.8 CAUSES AND RISK FACTORS

Our risk of developing an **AUD** may start before we're even born. Research suggests that a combination of genetics and environment influence the likelihood of misusing alcohol.[25] If someone's parents drank heavily, this may be a hereditable trait that they acquired from their parents, and it is suspected that in about 50% of cases this may be true. They may then pass on a tendency to drink to their offspring by modelling alcohol use as a coping strategy or just as a daily habit around their children. The combination of exposure to alcohol and having positive expectancies around alcohol *and* a genetic predisposition can further increase risk. Of course, this is rather an oversimplification. There will be many other mediating factors that determine the outcome for any individual. Often, a child's experiences of parental drinking make them determined not to follow the same path.

From an employer's perspective, it's important to remember that genetic factors are just one element of a broader picture. An alcohol-positive environment such as pro-alcohol workplace policies, practices and expectations can increase the likelihood of excessive consumption and subsequent problems. Usually, there will be an underlying trigger—stress, pressure, relationships difficulties, for example, and work may be a contributory factor. There are risk factors for alcohol misuse due to the nature of the work itself, so these need to be considered alongside the organisation's alcohol culture.

Individuals who have experienced traumatic events may use alcohol to self-medicate and numb emotional pain. Unresolved trauma can lead to post-traumatic stress disorders (PTSD), which are associated with higher rates of alcohol misuse, with research stating that up to 75% of military veterans with a diagnosis of PTSD also had a co-morbid alcohol or substance use disorder.[26] But this is not just a military issue, employees exposed to traumatic events at work, such as accidents, violence or witnessing death, are more likely to increase their alcohol consumption in the aftermath of the event. Also, the degree of alcohol use was correlated with the severity and frequency of the traumatic incidents, suggesting that the more traumatic the experience, the greater the risk of alcohol misuse.[27] Many rail companies have recognised this issue and have put in place support systems such as counselling, peer support programmes and psychological services to help drivers cope with the aftermath of traumatic events, such as witnessing suicides.

But it's not just trauma that can precipitate heavy drinking. High levels of stress at work or in personal life can lead individuals to turn to alcohol as a coping mechanism. The chronic activation of the stress response system can create a strong drive to self-medicate with alcohol. People with pre-existing mental health difficulties, such as depression, anxiety or social anxiety, may also turn to alcohol as a form of self-medication, i.e. to numb or avoid distressing symptoms. Alcohol's depressant effect can temporarily alleviate symptoms but exacerbate mental health problems in the long run.[28] Similarly, individuals with a history of substance abuse or addiction to other substances may be at a higher risk of developing AUDs. The transition from one addictive substance to another is not uncommon.

Fifteen per cent of people working in the private sector state that there is an expectation to drink alcohol at employer events, compared with 8% of public sector employees.

Private sector employees are 38% more likely to report workplace alcohol use than public sector.

Alcohol is 2.5 times more likely to be subsidised at work social events for private sector employees

(Drinkaware, 2022, Alcohol in the workplace)

3.8.1 Workplace Culture

An organisation's tolerance of alcohol use can influence employees' behaviour. Workplaces that promote social drinking or have a history of overlooking excessive alcohol use may inadvertently encourage risky drinking habits. Anecdotally, some industries have a greater boozy reputation than others. One study found that professionals working in London's financial sector were more likely to drink above recommended limits compared to the general population. This trend was attributed to the pressures of the job and the expectation to participate in social drinking events.[29]

In response to changing expectations in the workplace, Lloyds of London introduced a comprehensive zero-tolerance approach as part of their new alcohol policy in 2017. Initially, this applied only to staff, with brokers still being able to consume alcohol during the working day but two years later the policy was extended to everyone having access to Lloyds of London's buildings. But the move was controversial, and critics argued that traditions around alcohol use do not change simply by banning alcohol—a much more nuanced approach is necessary. The very fact that the ban was newsworthy across the world shows how embedded alcohol is in some institutions.

Even away from the traditional 'booze and shmooze' legacy of sealing deals over a drink, pro-alcohol attitudes are rife in many industries. Easy access to alcohol, such as in company-provided social events or a nearby off-licence, can contribute to alcohol problems. The more accessible alcohol is, the higher the risk of misuse. Influence from colleagues who encourage alcohol consumption can be a risk factor, too. Workplace peer pressure can lead employees to engage in drinking behaviours they might otherwise avoid. Lunchtime drinks, happy hours after work and an expectation, even if implicit, that employees will provide hospitality for clients or colleagues can put pressure on staff to drink.

QUESTIONS FOR REFLECTION

Consider your workplace

Is alcohol ever given as a reward, prize or gift to employees or customers?
Is alcohol available at work social events?
Is alcohol subsidised or provided free at work parties?
If you've answered yes to any of the aforementioned questions, what do you think would be the effects in the workplace of changing these practices and why?

3.9 THE IMPACT OF ALCOHOL ON WORK

The nature of an employee's role within the organisation can also influence how alcohol misuse manifests and its consequences. Frontline employees, such as customer service representatives or production workers, may experience immediate negative consequences due to impaired motor skills, coordination and response times. With almost one in four employees stating that they had experienced decreased productivity as a result of colleague's drinking,[30] this can have a marked effect on efficiency. Impaired functionality among frontline employees can pose safety risks in roles that require physical dexterity and vigilance such as assembly line work or driving jobs.

But it's not just at the frontline level that problems occur. Alcohol doesn't discriminate and people in managerial or executive roles are just as prone to alcohol problems as everyone else. Manager's actions also set an example for the rest of the workforce and beyond, so inappropriate drinking behaviour may also damage client relationships. Roles that involve complex cognitive tasks, such as financial analysis or legal work, are highly sensitive to the cognitive impairment caused by excessive drinking, with potentially disastrous implications for the organisation.

CASE STUDY

A 2018 employment tribunal revealed that a managing director struck a sales manager whilst under the influence of alcohol, resulting in a serious head injury. Although the incident didn't take place on work premises, vicarious liability was placed on the organisation.

(*Bellman v Northampton Recruitment Ltd* [2018] EWCA Civ 2214)

The increase in remote working during and after the COVID-19 pandemic may provide employees with more opportunities for discreet alcohol consumption during working hours. Researchers have found that isolation, homeworking and job insecurity during 2020 and beyond are correlated with changes in drinking patterns and around one in five UK consumers increased their consumption. In particular, it was those already drinking at risky levels who were most likely to drink more.[31]

Flexible work schedules afford employees opportunities to engage in drinking during traditional work hours, which can impact their ability to fulfil work-related commitments. This is a phenomenon identified from research into the impact of shift work on drinking patterns.[32] Understanding these factors is essential for HR and OH managers to tailor their interventions and support systems effectively. It highlights the need for a holistic and individualised approach to addressing alcohol misuse in the workplace, considering the unique circumstances of each employee. Luckily, this will be our focus for the next few chapters.

NOTES

1 UK Chief Medical Officers' Alcohol Guidelines Review Summary of the Proposed New Guidelines, 2016 (https://assets.publishing.service.gov.uk/media/5a7f51b4e5274a2e87db5206/summary.pdf).

2 Booker, L., 2018. *Try Dry: The Official Guide to a Month Off Booze.* London: Square Peg.

3 Office for Health Improvement and Disparities, 2024. Fingertips Alcohol Profile (https://fingertips.phe.org.uk/profile/local-alcohol-profiles).

4 Alcohol Change UK, 2023. Identifying and Addressing Cognitive Impairment in Dependent Drinkers (https://alcoholchange.org.uk/publication/the-blue-light-approach-identifying-and-addressing-cognitive-impairment-in-dependent-drinkers).

5 IARC Alcohol Consumption, 2010. Monographs on the Evaluation of Carcinogenic Risks to Humans Volume 96 Alcohol Consumption and Ethyl Carbamate (https://ar.iiarjournals.org/content/31/6/2402.3).

6 Rumgay, et al., 2021. Global Burden of Cancer in 2020 Attributable to Alcohol Consumption: A Population-Based Study. *The Lancet* (https://www.thelancet.com/article/S1470-2045(21)00279-5/fulltext).

7 Hydes, et al., 2019. A Comparison of Gender-Linked Population Cancer Risks Between Alcohol and Tobacco: How Many Cigarettes Are There in a Bottle of Wine? *BMC Public Health.* Hydes, et al., 2019. A Comparison of Gender-Linked Population Cancer Risks Between Alcohol and Tobacco: How Many Cigarettes Are There in a Bottle of Wine?

8 Cancer Research UK. How Does Alcohol Cause Cancer (https://www.cancerresearchuk.org/about-cancer/causes-of-cancer/alcohol-and-cancer/how-does-alcohol-cause-cancer#:~:text=Acetaldehyde%20can%20damage%20our%20cells,make%20cells%20divide%20more%20often).

9 NHS Inform Alcohol-related liver disease (https://www.nhsinform.scot/illnesses-and-conditions/stomach-liver-and-gastrointestinal-tract/alcohol-related-liver-disease/).

10 Adfam & Alcohol Concern UK, 2017. Supporting Family Members to Respond to Change Resistant Drinkers (https://adfam.org.uk/wp-content/uploads/2024/04/Blue-Light-Family-Project-Toolkit-Final.pdf).

11 Hernandez-Rubio, et al., 2022. Prevalence and Associations of Metabolic Syndrome in Patients with Alcohol Use Disorder. *Scientific Reports* (https://www.nature.com/articles/s41598-022-06010-3).

12 Bernadin, et al., 2014. Cognitive Impairments in Alcohol-Dependent Subjects. *Frontiers in Psychiatry* (https://www.frontiersin.org/journals/psychiatry/articles/10.3389/fpsyt.2014.00078/full).
13 Alzheimer's Disease International, 2014. World Alzheimer's Report (https://www.alzint.org/u/WorldAlzheimerReport2014.pdf).
14 Karadaiyan and Cutrera, 2013. Alcohol Hangover: Type and Time-Extension of Motor Function Impairments. *Behavioural Brain Research* (https://www.sciencedirect.com/science/article/abs/pii/S0166432813001782?via%3Dihub).
15 Scholey, et al., 2019. Effects of Alcohol Hangover on Cognitive Performance: Findings from a Field/Internet Mixed Methodology Study. *Journal of Clinical Medicine* (https://pmc.ncbi.nlm.nih.gov/articles/PMC6518120/).
16 Martin, et al., 1993. Development and Validation of the Biphasic Alcohol Effects Scale. *Alcohol Clinical & Experimental Research* 17(1), 140–146.
17 There is no published research on the impact of alcohol on people who are trans and whether or to what extent their body's response to alcohol changes during and after transitioning. For the purposes of this book, therefore, the terms 'male/man' and 'female/woman' will refer to gender assigned at birth.
18 Alcohol change UK What is Alcohol-Related Brain Damage? Factsheet (https://alcoholchange.org.uk/alcohol-facts/fact-sheets/alcohol-related-brain-damage-what-is-it).
19 White, A. M., 2020. Gender Differences in the Epidemiology of Alcohol Use and Related Harms in the United States. *Alcohol Research* Oct 29; 40(2), 01.
20 Women's Health Concern, Understanding the Risks of Breast Cancer. Poster (https://www.womens-health-concern.org/wp-content/uploads/2019/10/WHC-UnderstandingRisks-ofBreastCancer-MARCH2017.pdf).
21 However, members of Alcoholics Anonymous self-identify using the term 'alcoholic' and in this context it is purely descriptive, rather than judgemental.
22 International Classification of Diseases—10.
23 World Health Organization, 2018. Global Status Report on Alcohol and Health 2018 (https://www.who.int/publications/i/item/9789241565639).
24 Cornah, D., 2005. Cheers? Understanding the Relationship Between Alcohol and Mental Health. *Mental Health Foundation* (https://www.drugsandalcohol.ie/15771/).
25 National Institute on Alcohol Abuse and Alcoholism, Alcohol's Effects on Health Factsheet. Risk Factors: Varied Vulnerability to Alcohol-Related Harm (https://www.niaaa.nih.gov/health-professionals-communities/core-resource-on-alcohol/risk-factors-varied-vulnerability-alcohol-related-harm#pub-toc3).
26 Petrakis and Simpson, 2017. PTSD and alcohol Use Disorder; A critical Review of Pharmacological Treatments. Alcohol, Clinical and Experimental Research (https://pmc.ncbi.nlm.nih.gov/articles/PMC5375032/).
27 Bacharach, et al., 2008. Firefighters, Critical Incidents, and Drinking to Cope: The Adequacy of Unit-Level Performance Resources as a Source of Vulnerability and Protection. *Journal of Applied Psychology* (https://pubmed.ncbi.nlm.nih.gov/18211142/).
28 Robinson, et al., 2009. Self-Medication of Anxiety Disorders with Alcohol and Drugs: Results from a Nationally Representative Sample. *Journal of Anxiety Disorders* (https://www.sciencedirect.com/science/article/abs/pii/S0887618508000856).
29 Morris, et al., 2012. Insight into City Drinkers Insight into City Drinkers, Alcohol Use, Attitudes, and Options for Addressing Alcohol Misuse in the City of London (https://democracy.cityoflondon.gov.uk/documents/s7912/Insight%20into%20City%20Drinkers.pdf).
30 Alcohol Change UK, alcohol in the workplace (https://alcoholchange.org.uk/alcohol-facts/fact-sheets/alcohol-in-the-workplace).

31 Alcohol Change UK, 2020. Dinking During Lockdown: Headline Findings (https://alcoholchange.org.uk/blog/covid19-drinking-during-lockdown-headline-findings#:~:text=Six%20per%20cent%20have%20stopped,drinking%20more%20frequently%20under%20lockdown).
32 World Health Organization Europe, 2013. Status Report on Alcohol and Health in 35 European Countries (https://movendi.ngo/wp-content/uploads/2019/11/Status-Report-on-Alcohol-and-Health-in-35-European-Countries.pdf).

4 Conduct or Capability? *Recognising and Responding to Signs of Alcohol Problems*

All the while, I was somehow managing to hold it together and work as acting celebrity editor on Woman & Home magazine. I was still managing to pass as a regular person, sort of. I was producing decent interviews with wholesome celebrities . . .

I was like a house that looks presentable enough on the outside. But when you open the door, you're engulfed by a cloud of dust, unseen inhabitants scurry around, none of the lights work, and there's unopened mail all over the floor.

(Catherine Gray, The Unexpected Joy of Being Sober)

4.1 INTRODUCTION

In this chapter, we examine conduct and capability issues in the workplace, clarifying when an employee's actions breach company policies (conduct) and when their ability to carry out their contractual duties is compromised through a long-term alcohol problem (capability). We will also consider the early signs and symptoms of alcohol misuse to look out for.

The goal is to facilitate early identification when someone in your workplace may be struggling with their drinking. We will also explore numerous opportunities for early intervention, including screening for alcohol misuse, the use of questionnaires and assessment tools and strategies for identifying individuals who may require further assistance. Lastly, we will examine the profound impact of having a close loved one who is alcohol-dependent, encompassing financial, emotional, relational, spiritual and social dimensions.

QUESTIONS FOR REFLECTION

What signs would indicate to you that an employee may have an AUD?
What signs would indicate to you that an employee was intoxicated or hungover whilst at work?

DOI: 10.1201/9781032648927-5

As you can see from the aforementioned examples, we are talking here about two distinct problems in the workplace:

- Acute alcohol consumption (intoxication and hangovers) leading to conduct issues and
- The chronic effects of excessive drinking (AUDs) leading to capability, and potentially conduct, problems.

There's a lot of cross-over in the signs of each of these issues, so it's important to consider the pattern of these presentations—are they frequent? Are there signs from several different categories? Are there signs that are not explainable by any other cause?

The ability to distinguish between conduct and capability increases the likelihood that individuals receive the assistance they need while maintaining a safe and productive work environment. It also promotes an environment where staff are confident that they will be supported if they disclose an alcohol problem or related disorder.

Let's start by looking at some of the signs you might notice if an employee is under the influence of alcohol at work or if the consequences of heavy drinking over time are starting to show.

4.2 SIGNS AND SYMPTOMS OF ALCOHOL MISUSE

4.2.1 Physical Signs

We have already seen in Chapter 3 that alcohol can impact the whole body and, as everyone responds differently to alcohol, for some people even low levels of alcohol use can take their toll on that person's health. The overview of physical signs below is not an exhaustive list, nor does any one of these signs indicate an alcohol issue on its own. When trying to determine whether a colleague's symptoms may be alcohol-related, don't forget to weigh up other potential markers and consider behavioural and emotional indicators, risk factors and any other information that you may discover that point to alcohol rather than another cause.

- Flushed face often accompanied by redness or persistent blotchy skin can be a noticeable physical sign of alcohol consumption.
- The unmistakable smell of alcohol on the breath or clothing is a classic sign of recent alcohol consumption. When this odour is present during working hours, it may be an indication of a problem.
- Hand tremors or visible shakiness, particularly when not attributed to another medical condition, may suggest alcohol withdrawal symptoms. These physical manifestations can become more pronounced as an individual's alcohol problem progresses.
- Alcohol impairs motor skills and coordination. Slurred speech, an unsteady gait or difficulty with fine motor tasks like typing or handling objects can be signs of alcohol-related impairment, either through intoxication or as a result of long-term drinking
- Unexplained weight loss or gain, unrelated to a medical condition, can be associated with alcohol misuse. Alcohol can disrupt eating habits and affect nutrition.

4.2.2 Behavioural Signs

Behavioural changes related to drinking are often the first cause for concern and the most noticeable, affecting the employee's job performance and workplace relationships. Don't expect to see all of the signs mentioned here, likewise, there may be entirely different explanations for behaviour changes. Either way, you'll want to get to the bottom of what is impacting on your colleague in this way.

- Frequent absenteeism, lateness or unexplained and prolonged absences from work tasks can be indicative of alcohol-related problems. Someone struggling with alcohol misuse may find it challenging to maintain regular attendance.
- A noticeable decline in job performance or productivity, missed deadlines or increased errors can suggest alcohol-related impairment. An employee's ability to concentrate and complete tasks effectively may be compromised.
- Increasing interpersonal conflicts, arguments or confrontations with co-workers or supervisors can be a sign of alcohol-related behaviour changes. Alcohol misuse may lead to irritability and difficulty managing relationships.
- Individuals struggling with alcohol issues may withdraw from social interactions, exhibit reluctance to attend work-related events or distance themselves from colleagues. This isolation can be a protective measure to hide their alcohol use.
- A noticeable decline in personal grooming, hygiene and dress standards can be an external manifestation of alcohol misuse. Employees may appear dishevelled, unkempt or inappropriately dressed for the workplace.

4.2.3 Emotional Signs

Emotional changes related to alcohol misuse can drastically impact a person's well-being as well as work performance. As escalating alcohol problems often coincide with times of stress or other difficulties, it can be hard to distinguish between psychological distress due to circumstances or deteriorating mental health, and those precipitated by alcohol use.

- Frequent and extreme mood swings, including irritability, anger or sudden shifts in emotional state may be related to alcohol use. These mood fluctuations may disrupt workplace relationships and dynamics.
- Prolonged periods of sadness, hopelessness or changes in sleeping and eating patterns can be linked to alcohol-related emotional struggles. Low mood and depressive symptoms may also be indicators of excessive drinking.
- Excessive worry, restlessness or panic attacks may result from alcohol-induced anxiety or withdrawal symptoms. Often, alcohol is consumed in an attempt to relieve symptoms of anxiety or sleeplessness and may work well in the short-term but is likely to cause an escalation of symptoms over the longer term.

- Individuals struggling with alcohol issues may become defensive or evasive when asked about their drinking habits or work-related issues. This is often to conceal their own alcohol-related concerns.

TABLE 4.1
Checklist of Signs of Problematic Alcohol Use and How They May Present in the Workplace

Impaired job performance	**Cognitive impairments**
Decreased productivity and quality of work	Poor task recall, impaired working memory
Frequent errors or missed deadlines	Difficulties in paying attention, impulsivity
Inconsistent work patterns, e.g. periods of high productivity followed by periods of poor performance or absenteeism	Impaired planning, decision-making and prioritising
Change in work habits	Difficulty focusing on or completing a task
Reduced initiative or creativity	**Absences and lateness**
Different levels of impairment throughout the day	Consistent lateness or arriving at work in a dishevelled state
Erratic behaviour	Unexplained absences during the working day
Difficulty concentrating or staying focused on tasks	Frequent short-term absences, particularly on Mondays and/or Fridays
Lethargy, fatigue or drowsiness	**Change in appearance and hygiene**
Unprofessional conduct, such as inappropriate comments or actions	Neglect of personal grooming and hygiene, such as unwashed clothing, dishevelled appearance or poor personal care
Bringing alcohol onto premises or hiding alcohol in the workplace	Noticeable weight loss or gain
Excessive risk-taking	**Social isolation**
Sober-shaming or pressuring others to drink	Avoidance of social gatherings, both inside and outside the workplace
Frequent accidents in or outside of work	Secretive or evasive behaviour
Emotional signs	Withdrawal from workplace events or activities
Mood swings, irritability, aggression or emotional instability	Strained relationships with coworkers or isolation from colleagues
Unpredictable and inconsistent responses	**Communication problems**
Defensiveness about alcohol or work issues	Erratic communication, such as unresponsiveness to emails or phone calls
Physical signs	Conflict with colleagues
Slurred speech	Constant references to own drinking behaviours
Unsteady gait or poor coordination	Difficulty in maintaining clear and effective communication with colleagues, supervisors or clients
Bloodshot or glassy eyes	Erratic communication, such as unresponsiveness to emails or phone calls
Odour of alcohol on the breath or clothing	Unexplained discrepancies

(Continued)

TABLE 4.1 (*Continued*)
Checklist of Signs of Problematic Alcohol Use and How They May Present in the Workplace

Impaired job performance	**Cognitive impairments**
Odour of vomit on breath or clothing	Missing or misplaced items, documents or equipment
Tremors or shakiness	
Financial difficulties	Discrepancies in financial records or unaccounted for work hours
Sudden or unexplained financial problems	**Safety concerns**
Frequent requests for advances or loans from coworkers or the employer	Unsafe work behaviours or actions that pose a risk to personal safety or the safety of others
Decline in attendance at work-related events	Impaired coordination or motor skills
Decreased attendance/participation in work meetings, training or events	Inability to operate machinery or equipment safely due to impairment
Consistent avoidance of company-sponsored functions	**Health problems**
Other	Frequent illness or health issues, which may be related to alcohol misuse
Alcohol-related conviction e.g. drink-driving, drunk and disorderly	Stress-related health problems, such as hypertension or gastrointestinal issues
Co-worker observations	Deteriorating mental health including low mood, anxiety, social anxiety and depression
Loss of interest in previously enjoyed activities	Poor sleep
Normalising intoxication	

Despite this lengthy list, unless an employee states that they are having difficulties controlling their alcohol use or admits that alcohol is impacting their work, it can be very difficult to identify an issue. Even the most empathic employer is unlikely to find that employees willingly disclose alcohol misuse or even acknowledge it to themselves. This is why it's important to create both an ethos and a structure within the workplace that focuses on the wellbeing of employees. Policies are helpful, but it will be in everyday interactions, the behaviour of leaders and experiencing practical examples of support for colleagues affected, that will determine whether the workplace is a 'safe' place for disclosure.

If alcohol is being used as a coping strategy, as it so often is, it is important to tackle the underlying problem. Alcohol use may be attributed to work-related stress, bullying, burnout, rust-out or job insecurity, among others.

There may also be outside factors that an employee is responding to by drinking. These can include bereavement, divorce or separation, moving house or other stressful life events. Where employees are experiencing these events, you may wish to instigate extra support for a while. It's important to acknowledge that they may be using alcohol as a coping strategy and offer help to widen their repertoire of coping responses. It is often life's events that spark increased drinking and whilst we may be understanding of an employee's temporary reliance on alcohol, we can serve them best by pointing them in the direction of more positive ways of dealing with their difficulties.

4.3 HYBRID AND REMOTE WORKING

Traditional workplace settings allow managers and colleagues to observe behavioural cues such as changes in appearance, demeanour or attendance, that might signal alcohol misuse. With the move to remote work, these visual and social indicators are no longer readily available, making it more difficult for employers to detect early warning signs.

One of the most difficult challenges in a remote setup is reduced visibility. In an office environment, supervisors could notice signs of alcohol-related struggles through face-to-face interactions. Remote work, however, limits these spontaneous encounters, making it harder to observe subtle changes in employee behaviour. Performance issues like missed deadlines or inconsistent work quality can still be tracked, but they may be attributed to general stress or technical challenges rather than alcohol misuse.

Additionally, the isolation that often accompanies remote work can exacerbate alcohol-related issues. Without the regular social interactions of the office, employees may feel disconnected and resort to alcohol as a coping mechanism. This isolation can make it harder for colleagues or supervisors to provide support or intervene. The blurred boundaries between work and home life in remote settings also contribute to the problem. Employees working from home might be more inclined to drink during the workday, feeling that they are less likely to be monitored.

To address these challenges, you could implement several strategies. One approach is to increase the frequency of virtual check-ins. Regular video meetings not only help maintain team cohesion but also provide an opportunity for managers to gauge the wellbeing of their staff. These conversations can help identify shifts in mood, energy or engagement that might indicate underlying issues.

You could also employ digital tools to support employee wellbeing. Online counselling services, wellness apps and virtual workshops have become common offerings. These resources allow employees to access help discreetly and at their convenience. Training managers to recognise signs of distress in a remote context is helpful. Workshops and guidelines now emphasise the importance of paying attention to changes in communication patterns, productivity and engagement in virtual spaces.

In hybrid work environments, where employees alternate between home and the office, organisations have a unique opportunity to combine remote support with in-person check-ins. On-site days can include wellness programmes or individual meetings focused on employee health and wellbeing. This blended approach allows for a more comprehensive understanding of employees' needs and challenges.

Employers must also consider long-term strategies to address alcohol-related issues in remote and hybrid settings. Creating an open and supportive environment is essential. Encouraging discussions about mental health and substance use, offering anonymous feedback opportunities and normalising the use of support services can help reduce stigma and encourage employees to seek help. Technology can play a role in this process, not only in monitoring work patterns but also in providing personalised wellness resources and self-assessment tools.

Ultimately, while remote and hybrid work presents new challenges in managing alcohol-related issues, they also offer opportunities for innovation in support and intervention. Online tools and digital programmes offer a new way of accessing support, that may be more attractive to busy workers.

4.4 DEALING WITH IMPAIRED PERFORMANCE AND PRODUCTIVITY

As a supervisor or line manager, your knowledge about problems related to alcohol use is not nearly as important as your knowledge of the specific steps to take when an individual employee's work performance is poor or impaired by these problems.

(Tobutt, C., 2016, Alcohol at Work)

4.4.1 Conduct or Capability (Capacity)

A conduct issue arises when an employee's actions breach company policies, practices ethos due to alcohol misuse. This includes behaviours such as frequent unexplained absences or patterns of sick leave due to alcohol-related issues; disregarding company rules or refusing to follow instructions or engaging in disputes or conflicts with coworkers as a result of alcohol-related behaviour.

A capability issue arises when an employee's ability to perform their contractual duties is compromised due to alcohol misuse. This may include a noticeable decline in an employee's output and performance; an employee's inability to meet the expectations outlined in their job description or when an employee's alcohol misuse poses safety risks to themselves or others in the workplace.

Many employers have a capability procedure separate from their disciplinary procedure to clearly distinguish between the issues of conduct (an employee's behaviour at work) and capability (an employee's ability to do their job).

It is normally a conduct issue if the employee has control over their actions. But, as ACAS explains, 'usually it's a capability issue if the employee has no control over it. For example, if an employee becomes unable to do their job due to an illness or disability, and adjustments or support could not help'.

For example, absence on grounds of illness or injury may be dealt with through the sickness absence policy or capability procedure, but absence for no good reason may call for disciplinary action.

(Model Capability Procedure, 2023, UNISON)

It is always advisable to begin from the viewpoint of an alcohol problem as a *capability* issue, where possible. Treating an employee's difficulties as a health concern and providing every opportunity for support is likely to be the most effective route to change. Seldom do sanctions and reprimands provide a basis for improved employee-employer relations and are highly unlikely to precipitate a change in drinking behaviour for the better. Treating the consequences of alcohol misuse as a deficit in capability, just like any other health issue that can reduce performance, fosters a supportive environment. This creates an atmosphere of trust and increases the likelihood that you can start open conversations about alcohol that lead to positive conclusions. Nevertheless, where there are significant or repeated infringements, a robust response to poor conduct is indicated. Drinking in the workplace is a health, safety and wellbeing issue and despite health and safety being the responsibility of every employee, the law firmly places legal onus on the employer to take timely action.

4.4.2 The Cycle of Change

Prochaska and DiClemente's Transtheoretical Model of Change,[1] often referred to as the Cycle of Change Model, offers a framework for understanding how individuals navigate behavioural change, particularly in areas like substance use. This model posits that change is not a singular event but a process involving distinct stages. Each stage reflects a different level of readiness and commitment to altering behaviour, which is fundamental to understanding and supporting employees who may struggle with alcohol use.

The stages begin with precontemplation, where individuals are unaware or in denial of the need to change. Employees in this stage may not recognise that their drinking behaviours are problematic or negatively affecting their health, performance or relationships at work. Awareness campaigns and nonjudgemental dialogue can be particularly valuable in this phase, as they help employees consider the potential consequences of their actions.

In the contemplation stage, individuals acknowledge that their behaviour might be causing them some difficulties and start weighing the pros and cons of change. For employees, this may manifest as recognising that their alcohol use is affecting their productivity, attendance or workplace interactions. However, ambivalence often marks this stage, as individuals are unsure whether they are ready or able to commit to change. Employers can play a supportive role here by providing access to educational resources or offering confidential counselling services to help employees explore their options.

The preparation stage involves a decision to change and the development of a plan to do so. Employees may seek out resources, such as EAPs or discuss their goals with trusted colleagues or supervisors. Providing a clear pathway to support—such as workshops, peer groups or external treatment options—can make a positive difference during this phase.

During action, individuals actively work on changing their behaviour, implementing the strategies they've developed. For an employee reducing or eliminating alcohol use, this might involve attending support meetings, adopting healthier coping mechanisms or setting clear boundaries around drinking. Employers can bolster this effort by offering workplace accommodations, such as flexible schedules or reduced workloads during high-stress periods.

The maintenance stage focuses on sustaining the change over time and avoiding relapse. For employees, this requires ongoing commitment and support to prevent old habits from resurfacing, particularly during stressful situations. Employers can reinforce maintenance by ensuring that the workplace environment concerning alcohol does not present opportunities for relapse, and by offering ongoing resources and encouragement.

Finally, the model acknowledges the possibility of relapse as part of the process. Note that this is not failure, but just another stage in the cycle. Employees may revert to previous behaviours temporarily, but with the right support, they can re-enter the cycle of change. Understanding that relapse is a common part of recovery helps reduce stigma and encourages employees to keep striving towards their goals.

The Transtheoretical Model is highly relevant in addressing employee drinking because it recognises that change is a personal journey, often requiring tailored interventions at different stages. By aligning workplace policies and support systems with this model, you can create a compassionate framework that meets all employees where they are, empowering them to address alcohol-related challenges in a constructive and non-judgemental way. You will find information on appropriate interventions and resources for each stage of the cycle in Chapter 9.

4.4.3 Severity of Alcohol Problems

Alcohol-related problems can vary in severity, and it's essential to recognise this spectrum when addressing an employee's capability and conduct. Tackling the issue at an early stage can prevent escalation to serious problems both for the individual employee and the organisation. Here is a breakdown of the levels of severity and their potential impact on an employee.

> *Even functioning alcoholics can still achieve great things and hold down jobs, (I invented a product and brought it to market during the years that my drinking progressed . . .)*
>
> *In the later years, however, it is alarming how rapidly things escalate and how the events fuelled by alcohol go from being mildly embarrassing or upsetting to perhaps dangerous or catastrophic, and how the periods of time in between these catastrophic events become smaller and smaller until, ultimately, alcohol is running the show completely and normal life disintegrates.*
>
> (From Rock Bottom to Sober Forever, Susan Laurie)

It's always best to tackle workplace alcohol issues head on. Left unchallenged, alcohol problems are unlikely to resolve themselves. By creating an open and transparent environment, those who need help feel more confident to ask for it, those in charge will be confident in their responses and colleagues affected will both be able to speak up about the impact and respond with compassion.

TABLE 4.2
A Summary of the Severity of Alcohol Problems and Their Potential Impact on the Individual in the Workplace

	Mild	Moderate	Severe	Crisis
Features	Mild alcohol-related issues may involve occasional social drinking or infrequent episodes of excessive drinking.	Moderate alcohol-related problems typically involve regular heavy but not constant alcohol misuse. Low mood, lethargy and lapses in memory or insight are possible	Severe alcohol-related problems often indicate chronic and heavy alcohol misuse, potentially leading to addiction or dependence. Short-term memory and physical dexterity may be compromised. There may be health consequences such as hypertension or anxiety	Crisis situations represent the most severe end of the spectrum, where an employee's alcohol misuse has reached a critical point. They may be drinking or frequently impaired during working hours. Decision-making skills and judgement are likely to be impaired. Physical and mental health impacts are likely
Impact on capability	There may be a minimal impact on an employee's capacity to perform their job duties, but they can still result in occasional absenteeism or decreased focus.	Employees may experience more frequent absenteeism, particularly short absence, reduced productivity and difficulty meeting deadlines.	Alcohol can severely impair an employee's capacity to perform their job duties. They may struggle with consistent attendance, experience significant decreases or fluctuations in productivity and struggle to meet job requirements.	In crisis situations, an employee's capacity to perform their job duties is severely compromised. They may be frequently absent, unable to function at work or incapable of meeting basic job requirements. They may disappear for periods during the working day.
Impact on conduct	Conduct may be affected in the form of occasional tardiness or minor interpersonal conflicts	Conduct issues may manifest as mood swings, strained relationships with colleagues and an increased likelihood of conflicts in the workplace.	Employees with severe issues may exhibit disruptive behaviour, including aggression, erratic emotional states and conflicts with coworkers or supervisors.	Conduct issues can be highly disruptive, including outbursts or lack of appropriate restraint, severe conflicts and potentially unsafe behaviour in the workplace.

HR and OH managers, supervisors and colleagues can facilitate early intervention through various strategies:

1. Develop a bespoke alcohol policy that sets down both the rules around alcohol consumption in your organisation and the implications of any breach. It should also outline the measures that will be taken to support an employee with an AUD. See Chapter 5 for further information. It should also dovetail with your other policies.
2. Implement a screening process to identify who is at risk of alcohol misuse. Confidential self-assessment questionnaires, discussions with HR or alcohol screening tools can be effective methods. For example, you could encourage employees to complete an anonymous alcohol screening questionnaire as part of their annual health assessment.
3. Train supervisors and managers to observe and document changes in behaviour, attendance and performance that may be indicative of alcohol-related issues. Proper documentation helps build a case for intervention and support. See Chapter 9 for details.
4. Create an atmosphere of open communication where employees feel comfortable discussing their concerns about a colleague's alcohol use. Encourage colleagues to express their worries in a non-judgemental and compassionate manner.
5. Develop a system to refer individuals showing signs of alcohol misuse to appropriate resources or external treatment providers. Timely referrals can make a huge difference in an employee's recovery journey. A list of trusted treatment providers and helplines could form an appendix in the alcohol policy, for example.

4.4.4 Performance Issues

Changes in employee performance that have been attributed to alcohol will generally fall under the province of capability issues. I cannot overstate that responding to problems at an early stage is the key to effective and timely performance management.

It's a good idea to implement performance metrics to track and address performance issues, offering support and intervention when necessary. Implementing performance improvement plans (PIPs) for employees dealing with alcohol-related performance challenges should take a supportive and empathetic approach. Next is a step-by-step process to establishing the effectiveness of such plans. Although, of course, you may already have processes in place to deal with performance issues related to other causes and these should adapt easily to this topic.

Throughout the entire process, maintain a compassionate and non-judgemental approach. Emphasise that the goal is to support the employee in overcoming their challenges and returning to optimal job performance while addressing their alcohol-related issues.

CASE STUDY

The story of Claire (not her real name), a successful bank executive and an Alcohol Change UK Community Champion, highlights the issue of 'grey area drinking' and how organisations can help. The 'grey area' lies between social drinking and addiction. It's ambiguous and amorphous and can disguise itself as a good time. But it is also a high wire act that leaves the drinker feeling permanently tired, anxious and in need of another drink to 'feel better'.

Claire recognised the signs and sought help from her employers, who she thought would be supportive. They were. Not only did they help Claire to get back on track, but also later, when she had given up alcohol completely, Claire approached her manager about an idea to offer support to colleagues. She was given time and funds to develop a sober support group that has helped many colleagues and inspired other UK businesses. She's even spoken at Whitehall, campaigning for changes to the laws around alcohol dependence in the workplace.[2]

4.5 ALCOHOL SCREENING AND BRIEF ADVICE

4.5.1 Alcohol Screening

Screening your workforce for alcohol problems offers several benefits: it's cost-effective, it identifies problems at an early stage, and it can have a role in preventing the development of AUDs. What's more, it's a highly accurate method of determining who is drinking at high-risk levels. In other words, it's not aimed at identifying dependent drinking but targets 'social drinkers'—those who drink normally by society's standards—whose drinking may be unwittingly putting them at risk of harm. As over 10 million people in the UK (around 14% of the adult population) regularly exceed the government's recommended weekly maximum of fourteen units, it's likely that a large minority of employees fall into this category. Screening is also a useful means of opening a conversation about alcohol in a workplace setting. There's a wealth of evidence that screening can be conducted effectively by line managers, wellbeing champions and HR personnel, with just a little training.

Screening is advisable as a preventative measure, although it can be used in response to alcohol-related workplace incidents. It is particularly useful as a wellbeing tool to support employees and to promote *capability* rather than as a result of *conduct* issues. Screening can provide a profound teachable moment, particularly when implemented as part of a wider wellbeing-centred approach to alcohol misuse.

4.5.2 Screening for Alcohol Use

One of the most used and best-evaluated alcohol screening tools is a ten-question questionnaire giving guidelines on scores that should precipitate brief advice or referral to further support. It takes a few minutes to administer and there is evidence to show that this and other screens[3] are sensitive enough to identify hazardous

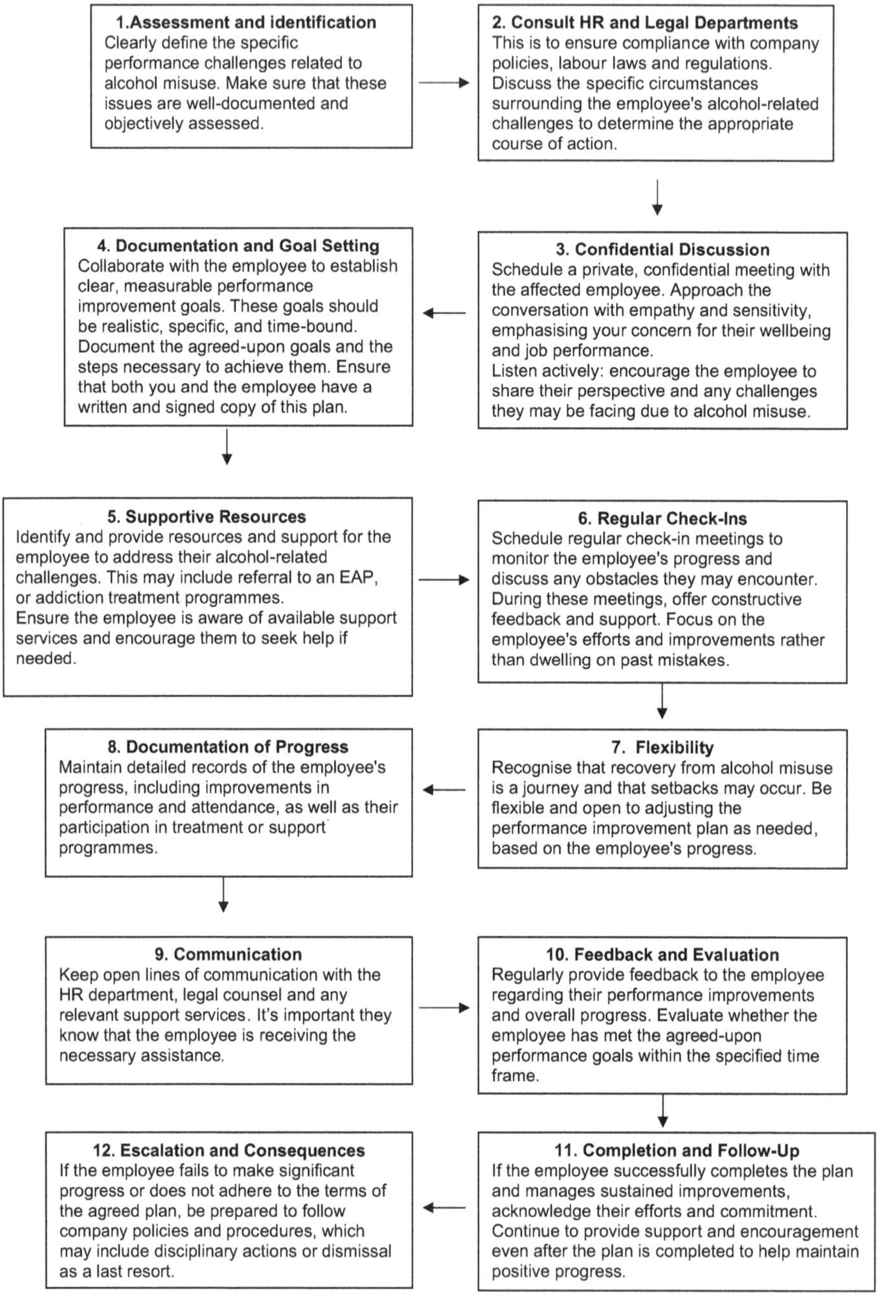

FIGURE 4.1 Case study

and harmful drinking, and subsequent brief interventions can then be delivered in response to high screening results.

Based on estimates of cost-effectiveness and preventable disease burden, alcohol screening and brief intervention ranks higher than other common preventive practices such as screening for hypertension, high cholesterol and cancer . . .

. . . As with any other health condition, it's better to prevent alcohol problems from developing or to treat them before they become severe.[4]

AUDIT stands for Alcohol Use Disorders Identification Test. It was developed by the WHO and is advised in UK national guidance as the recommended alcohol screening tool to use to determine the nature and severity of alcohol misuse.[5]

During your relationship with an employee, there may be many opportunities to introduce the **AUDIT** tool. It can be used in a wide variety of settings. Embedding regular screening into normal work practice can be reassuring to staff—everyone gets asked at some point, no-one feels singled out. You may wish to incorporate the tool into your standard paperwork as a wellbeing initiative or during the rollout of a new alcohol policy or procedures. You could also implement screening as part of a return- to-work interview, in response to sickness absence, after a change in work performance or behaviour or when there is suspicion of inappropriate alcohol use.

It can be difficult to know how to start a conversation about someone's drinking, but it's best to put the focus *on the harm* rather than *on the alcohol.* For example:

'You seem stressed lately, and I wonder how you're coping with that'.
'You don't seem yourself and I'd like to find out what's going on and how we can help'.

THE AUDIT COMPRISES THREE DOMAINS THAT EXPLORE DIFFERENT ASPECTS OF ALCOHOL HARM

The questions can be broken down into three domains:

Questions 1–3 These focus on levels and frequency of drinking and help to determine increasing risk use. These questions are where individuals are most likely to score points. This domain highlights the extent and frequency of drinking and is an excellent opportunity to educate the recipient about alcohol units, recommended limits and risk.

Questions 4–6 These questions look at early signs of dependent drinking and individuals tend to only score on these questions if they are also scoring on questions 1–3 and 7–10. This section shows how alcohol may be impacting on normal functioning and impairing the individual's ability to perform.

Questions 7–10 The final questions look at high-risk drinking and individuals may score points here if they are also scoring on the first 3 questions. These questions reveal ways in which alcohol use may disrupt everyday life and impact on others around the individual.

AUDIT is easy to score; you just add up the points to give a total up to a maximum of 40. This tool is most sensitive in the 0–20 range as it is designed to look at early signs of alcohol problems. It cannot be used to diagnose alcohol dependence. A score of 20+ indicates that the individual is experiencing substantial issues as a result of their drinking and would benefit from additional support to make changes.

Score 0–7 low risk. This score indicates that the individual is generally drinking within safe guidelines.

Score 8–15 increasing risk. This applies to anyone drinking over fourteen units per week, that hasn't observed any alcohol-related problems. Men drinking eight plus units a day and six plus units a day in women are also at increased risk of harm even though some may not exceed the weekly maximum. People in this category will not usually seek support, although some may realise their drinking could put them at risk. They may be hungover at work from time to time, experience short sickness absences or suffer low moods or trouble sleeping.

Score 16–19 high- risk. People scoring in this category are likely to have noticed some negative consequences of their drinking but may not have made a connection between their drinking habits and associated issues. As with the previous category, they may experience low mood and increased short absences but may also show poor timekeeping or performance or experience adverse impacts on relationships and physical health. If the employee wants support, you may wish to suggest a referral to your EAP or other specialist services and encourage them to come back to you if they have concerns.

Score 20–40 potential dependence. This score refers to drinking associated with significant difficulties. A quick reminder though: **AUDIT** cannot be used to diagnose dependence. However, people in this category have usually also experienced alcohol-related problems but may attribute negative consequences to other causes. Those with higher scores (30+) will have high levels of tolerance to alcohol and *may* show signs of physical withdrawal if they stop drinking. They may be drinking before or during work hours and their reliance on alcohol is likely to impact on their performance and behaviour at work. Does employee want support? If so, you can make a referral to appropriate services as given earlier. You will find information about alcohol treatment and recovery services in Chapter 8.

According to figures published by the Ministry of Defence,[6] 61% of personnel scored five or more on an alcohol screening test (AUDIT-C, the first three questions of AUDIT), indicating at least increased risk of alcohol-related harm.

4.5.3 The AUDIT Tool

This tool can also be found on the following website where it can be completed anonymously: www.alcoholchange.org.uk.

TABLE 4.3
The Alcohol Use Disorders Identification Test (AUDIT); an Evidence-Based Health Screen to Identify High-Risk Drinking

		Scoring system					
Domain	**Questions**	**0**	**1**	**2**	**3**	**4**	**Your score**
1	How often do you have a drink containing alcohol?	Never	Monthly or less	2–4 times per month	2–3 times per week	4+ times per week	
	How many units of alcohol do you drink on a typical day when you are drinking?	1–2	3–4	5–6	7–9	10+	
	How often have you had 6 or more units if female, or 8 or more if male on a single occasion in the last year?	Never	Less than monthly	Monthly	Weekly	Daily or almost daily	
2	How often during the last year have you found that you were not able to stop drinking once you had started?	Never	Less than monthly	Monthly	Weekly	Daily or almost daily	
	How often during the last year have you failed to do what was normally expected from you because of your drinking?	Never	Less than monthly	Monthly	Weekly	Daily or almost daily	
	How often during the last year have you needed an alcoholic drink in the morning to get yourself going after a heavy drinking session?	Never	Less than monthly	Monthly	Weekly	Daily or almost daily	
3	How often during the last year have you had a feeling of guilt or remorse after drinking?	Never	Less than monthly	Monthly	Weekly	Daily or almost daily	
	How often during the last year have you been unable to remember what happened the night before because you had been drinking?	Never	Less than monthly	Monthly	Weekly	Daily or almost daily	
	Have you or somebody else been injured as a result of your drinking?	No		Yes, but not in the last year		Yes, in the last year	
	Has a relative or friend, doctor or other health worker been concerned about your drinking or suggested that you cut down?	No		Yes, but not in the last year		Yes, in the last year	
						Total	

4.5.4 Brief Advice

Evidence indicates that short motivational conversations about alcohol use, known as *brief advice,* are effective in encouraging people to reduce their alcohol consumption. These conversations can be useful for helping employees to initiate change, as well as re-affirming the benefits of low-risk drinking. Brief advice uses a series of short sentences that urge the individual to reflect on how their drinking may impact their life and offers suggestions to reduce potential harm. Whilst brief advice is an effective stand-alone tool and indeed can be used to elicit change in several health-related behaviours, it's a natural partner to alcohol screening.

4.5.5 Delivering Brief Advice

The term 'brief advice' covers a range of advice-giving interactions, generally lasting for 5–10 minutes. Opportunistic brief advice as defined here is not to be confused with specialist *brief interventions* delivered over several sessions in treatment settings. The WHO outlines a framework for the delivery of brief advice. This uses the FRAMES model, giving guidelines on the content of the conversation. FRAMES is the acronym for six elements or sentences which can be structured and tailored to the individual's AUDIT score and the questions where they have scored the most points.

Brief Advice is evidence-based and statistically, one in eight recipients will make changes to their drinking, but it needs to be tailored to the level of drinking and tailored to alcohol-related harm to be effective. And brief advice if just that—a short motivational conversation. It is *not* a treatment and is most effective in those who are not showing signs of dependence. If, during the course of conversations between manager and employee, it becomes apparent that an employee is drinking problematically, motivational techniques can be used to encourage referral to specialist agencies.

Along with basic information about alcohol, such as guidance on recommended limits or how to work out units, simple brief advice should include the following:

Structured, personalised **Feedback** on risk and harm, focusing on tangible impacts

Emphasis on the person's personal **Responsibility** for change (although support is available)

Clear **Advice** to the employee to make a change to their drinking

A **Menu** of options for how to make a change in behaviour

An **Empathic** and non-judgemental statement to show understanding of the employee's situation

Self-efficacy A positive statement to increase the person's confidence in being able to change their behaviour.

Brief advice is a useful tool for managers and supervisors as it provides an opportunity to raise the issue using an objective, standardised, evidence-based measure. It avoids provocative language and is an informal opportunity to pre-empt any issues and creates an atmosphere that encourages openness, enabling staff to tackle drinking levels before a problem develops. Several organisations offer training on delivering the AUDIT tool and brief advice for managers, supervisors and wellbeing champions.

CASE STUDY

A large telecommunications company hosted an alcohol awareness roadshow as part of their wellbeing calendar. Local alcohol treatment providers were invited to set up stalls in the foyer of the company's buildings for a day. Various activities were offered including:

- the wearing of 'beer goggles'—a tool for demonstrating the level of impairment at the drink-drive limit,
- a chance to explore how much liquid constitutes a single unit of alcohol, depending on the strength (%ABV) of the drink and how many units in a 'standard' drink
- free unit measuring cups and leaflets
- an invitation to sign up for the Dry January campaign

Employees were also offered five to ten minutes of private and confidential consultation with an alcohol recovery practitioner and receive screening and brief advice, after which they could be referred to further support if they so wished. The company was given information about numbers of employees that took up this offer and how many fell into risky categories, but anonymity was maintained for all employees. As a result of this highly successful initiative, the organisation turned this into an annual event and trained their wellbeing champions to also screen and offer brief advice.

4.6 HAVING A CLOSE LOVED ONE WITH ALCOHOL DEPENDENCE

4.6.1 The Impact of Having a Close Loved One with Alcohol Dependence

The impact of caring for someone who has a dependence on alcohol extends beyond the individual struggling with alcohol misuse. The financial toll can be devastating, and families may find themselves facing unexpected expenses and financial instability. Often though, it's the emotional toll of coping with a loved one's drinking that has the most impact. This can lead to stress, anxiety, depression and a sense of helplessness. Alcohol dependence can strain relationships with spouses, partners, children, friends and extended family members. Frequent conflicts, breakdowns in communication and emotional distance can spill over into the workplace. Families may withdraw from social activities to avoid exposing the loved one's alcohol issues. This social isolation can lead to feelings of loneliness and detachment from social networks and impact workplace engagement and productivity.

With almost one in three adults in the UK being affected by someone else's alcohol problem,[7] and naturally these difficulties are likely to impact work performance. Disclosing that someone you care about is struggling with addiction is challenging. There's still a great deal of stigma around alcohol problems and a lack of understanding of how loved ones are affected, so work may be a haven away from a chaotic home life.

Don't judge. We may think we know how we would respond to a loved one's alcohol problem, but addiction is insidious, it can develop slowly and consume every aspect of family life. There are no simple solutions so if you think you've got some great advice about what a colleague should or shouldn't do about the issue consider that they're likely to have considered all possible responses and outcomes and however they respond, they're doing the best they can in difficult circumstances. What will be welcomed is a non-judgemental listening ear, validation of their experience and the reassurance of your support. You may wish to signpost them to additional resources.

Most family members that I've worked with are looking for ways to change their loved one's drinking behaviour and reject the idea that they would benefit making changes or accessing support for themselves. Whilst it's understandable that this is the case, there is nothing that anyone can do to change someone else's drinking—not begging, bribing, shouting, leaving or pouring all the booze down the sink. But what they *can* change, is their resilience and response to what seems like a no-win situation. Counselling and group support are very effective in helping families to take practical steps to reduce the impact of another's AUD and meet their own needs.

There are over 20,000 alcohol-related deaths each year in the UK and grief from the loss of someone you love who was in active addiction or in recovery can be complicated. Either the manner of death or the nature of the relationship with the deceased can bring up feelings of stigma and shame which may require support to resolve, so if you have an employee who is faced with such a tragedy, your empathy and compassion will be valued.

4.6.2 Resources for Those Caring for a Loved One with an Alcohol Problem

Appropriate support can make a huge difference to families. Access to resources and guidance can alleviate the challenges they face. If your employee is dealing with alcohol issues at home, here are various resources that can provide assistance:

Adfam The national charity supporting anyone affected by a loved one's substance use. They offer counselling, group and individual support and have a vast collection of resources and research on this topic.

Al-Anon and Alateen These support groups offer help to family members and friends of individuals with alcohol problems. They provide a safe and empathetic space for sharing experiences and finding emotional support.

Community alcohol services Local community treatment may provide resources, counselling and support groups specifically designed for families affected by alcohol dependence. These resources can help families connect with others facing similar challenges.

Cruse A bereavement charity that offers specialist support to anyone bereaved by drugs or alcohol.

Kinship A national organisation supporting anyone who is caring for a loved one's children due to parental drug or alcohol use.

National Association for Children of Alcoholics (NACOA) offers support and information for anyone who grew up in a household with someone drinking problematically.

Therapy and Counselling Individual or family therapy can help people caring for a loved one with an alcohol problem develop coping strategies, improve communication and navigate the emotional complexities of their situation. The British Association of Counselling Practitioners (BACP) holds a list of qualified counsellors and psychotherapists.

NOTES

1 Prochaska, J. O. and DiClemente, C. C., 1984. Toward a Comprehensive Model of Change, in *Chapter, Treating Addictive Behaviours*, Eds. Miller, W. R. and Heather, N. Springer.
2 British Association of Counsellors and Psychotherapists Article, 2023. Alcohol at Work: A Heady Cocktail, BACP Workplace Magazine (https://www.bacp.co.uk/bacp-journals/bacp-workplace/2023/july/alcohol-at-work/).
3 Specific, Measurable, Achievable, Relevant and Time-bound such as CAGE and AUDIT-C.
4 National Institute on Alcohol Abuse and Alcoholism The Healthcare Professional's Core Resource on Alcohol, Screen & Assess: Use Quick, Effective Methods. (https://www.niaaa.nih.gov/health-professionals-communities/core-resource-on-alcohol/screen-and-assess-use-quick-effective-methods).
5 Alcohol-use disorders: diagnosis, assessment and management of harmful drinking (high-risk drinking) and alcohol dependence, 2011, National Institute for Health and Care Excellence.
6 Ministry of Defence, 2017. Alcohol usage in the UK armed forces: 1 June 2016–31 May 2017 (https://assets.publishing.service.gov.uk/media/5a8239e340f0b62305b93269/20170718_Alcohol_Usage_bulletin__-_O.pdf).
7 Adfam, 2016. Making it Happen (https://adfam.org.uk/wp-content/uploads/2024/06/Making_it_happen_final_PDF.pdf).

5 Developing an Alcohol Policy

Best Practice for Employers

5.1 INTRODUCTION

In this chapter, you will find a step-by-step guide for HR professionals on developing and implementing an alcohol policy. It emphasises the significance of clear communication, employee involvement and stakeholder engagement. Additionally, this chapter offers practical advice on auditing the workplace alcohol culture and includes guidance on recruitment practices and anti-discriminatory measures. A policy will not work in isolation, though. It needs to be part of a raft of measures aimed at improving the health and wellbeing of staff and reducing the harm caused by alcohol.

QUESTION FOR REFLECTION

To what extent is an employee's drinking the business of their employer?

If an individual's alcohol consumption isn't impacting on their work performance or relationships, and there's no suspicion of alcohol use or intoxication during work time, is it any concern of their employer what their drinking habits are, even if they are alcohol dependent?

Aside of safety-critical roles or other instances where alcohol-fuelled impairment may cause problems, is someone's drinking a legitimate target for workplace wellbeing or health initiatives?

Does the employer have a duty to educate and inform employees about behaviours that may affect their wellbeing and *could* impact on their work at a later stage?

Your answers to these questions should help you to determine the direction of your alcohol strategy going forward. Do you want to focus on reducing workplace risk or promoting employee wellbeing or both?

5.2 STEP 1: ASSESS THE NEED FOR AN ALCOHOL POLICY

Before embarking on the process of creating an alcohol policy, you'll need to determine your objectives and assess how it will benefit the organisation. Your policy should be the cornerstone of your response to problem drinking. It sets out the rationale for your procedures, details rules, roles and responsibilities, and indicates the consequences of infringement. It should also dovetail with other policies to provide a consistent and coherent response to problems.

DOI: 10.1201/9781032648927-6

More than that, it should embody the organisation's ethos around support for employees with difficulties and identify avenues of support as well as setting boundaries and disciplinary measures. An alcohol policy should be as individual as the organisation that develops it. It should be a dynamic document, ideally preventative as well as responsive, and regularly updated to reflect changes in the organisation. Your policy should also be coproduced by those that it affects.

According to a report by the Chartered Institute for Personnel Development, around one in five employers have dismissed an employee in the last two years, where drugs or alcohol misuse constituted as reason for the dismissal, and over one in three employers have disciplined an employee for alcohol misuse.[1] So, it's important to get this right.

The first step in policy development involves identifying what problem you're trying to solve, so here is a key question for you to reflect on:

What are your reasons for implementing an alcohol policy?

- To meet industry standards
- To support health and safety in the workplace
- To provide clarity around roles and responsibilities
- To minimise the impact of alcohol on productivity, performance and absenteeism
- In response to an incident
- Out of concern for staff wellbeing
- To address suspicions about employee drinking
- To establish the organisation's ethos and commitment
- When business practices change (e.g. at the implementation of workplace testing)
- Or for other reasons

Reflecting on your reasons for doing this work will help you to design the content of your policy and to determine its effectiveness. It's worth asking colleagues to answer this question too. There may be multiple reasons or multiple problems to address so if you're going to produce an impactful document, you'll need to know why you're doing it.

Here are some examples of UK organisations that have implemented alcohol policies to effectively address and prevent alcohol-related issues in the workplace:

British Airways The airline has adopted a stringent alcohol policy that includes regular training for employees, especially those in safety-sensitive positions. By establishing a zero-tolerance policy and providing support for those facing alcohol-related challenges, they have seen a notable decrease in incidents connected to alcohol consumption.

Network Rail NR has developed a comprehensive policy on alcohol and drug use, which includes regular testing for employees in safety-critical roles. This strategy not only helps to identify potential issues early but also offers support for those who need it, leading to improved safety records and employee health.

Greene King The brewery promotes responsible drinking through its policy, which emphasises mental health and wellbeing, leading to an environment where employees feel safe discussing alcohol-related matters.

NHS Trusts Various NHS organisations have established policies focusing on minimising alcohol-related harm among staff and students. These policies include education about the effects of alcohol and guidelines for responsible drinking at social events, contributing to a safer environment for both employees and patients.

These examples demonstrate how effective alcohol policies can enhance workplace safety, promote employee wellbeing and protect the organisation's reputation.

5.2.1 Risk Assessment

Alongside a policy, it is often helpful to make a risk assessment of the impact of alcohol on your organisation. What are the specific risks associated with drinking in your workplace? Do they include as safety hazards, decreased productivity, reputational damage, wellbeing or potential legal liabilities? What have you identified and what do you suspect?

How would you quantify these risks? What data do you have that can evidence the need for action? Alternatively, what data do you need to collect and how will you do this? Who needs to support this action and are they onboard with the changes that you want to make? Pushback from management or staff can lead to difficulties in implementing your policy and reduce its efficacy. How will you resolve differences in opinion?

Be prepared to consider how the work environment may actively contribute to problems, either as a result of workplace environment—pub culture, 'play hard work hard' ethos, corporate hospitality and normalised behaviours—or as a by-product of the working environment (see Chapter 1 for key workplace risk factors). If you can make an honest appraisal of the difficulties you're facing, your policy will be much more effective in reducing key risks.

Employers are legally obliged to that workplace policies on alcohol are fair and non-discriminatory. A blanket 'zero-tolerance' approach to alcohol misuse without considering reasonable accommodations for those with disabilities related to alcohol problems could be viewed as discriminatory. Policies should focus on health and safety considerations, taking individual circumstances into account and ensuring the employee has a fair opportunity to discuss health concerns.

5.2.2 Legal Requirements

Familiarise yourself with relevant laws and regulations, which may mandate alcohol policies or impact their content. Whilst employment law does not specifically prohibit alcohol use in the workplace, there may be industry requirements, particularly in safety-critical industries (such as construction) or roles (e.g. airline pilot) that require a robust response to alcohol as a potential hazard.

If you have a good understanding of the company's legal responsibilities you will be able to ascertain whether your policy is fit for purpose. Litigation in this area can be costly, both for the organisation and the individuals involved, so it's wise to ensure that you are aware of good practice in policy development in your industry. Information about key relevant UK legislation can be found in Chapter 7.

> During the COVID-19 epidemic, an employee with an unblemished record was dismissed for gross misconduct under Thameslink's 'zero-tolerance' alcohol policy. He had admitted to drinking a glass of wine before being called to cover a 'home standby' shift for a colleague. He had believed that he would not be called to work on that day and immediately stated that he'd consumed alcohol and would therefore not be fit for work. The dismissal was overturned as the disciplinary process did not consider that he had not worked the day in question and therefore presented no safety risk. The organisation's COVID-19 policy stated that the implication of not attending for work on a home standby day was to be marked absent and forfeit a day's pay.
>
> (*Wright v Govia Thameslink Railway*, 2022)

5.2.3 Stakeholder Input

Gather opinions from employees, management and other relevant stakeholders to understand their concerns and perspectives regarding alcohol use in the workplace. Your organisation's current attitudes towards drinking will need to be considered. Are staff in agreement with the tone and content of the policy? Is work needed to help staff to understand the importance of the measures you're taking? Consulting with staff at an early stage can forestall later problems. It can also be helpful to consider how the organisation's stance on alcohol might influence its public image and relationships with clients, suppliers and the wider community.

5.2.4 Conduct a Cultural Audit

Give some thought to how you're going to assess the current climate of alcohol use within the organisation. You could distribute anonymous surveys to gauge employees' perceptions of alcohol use in the workplace, their comfort levels and any observed issues (see the employee survey template given next). This will help to provide an evidence base for the need for a policy as well as starting a conversation in the workplace about alcohol.

Review past incidents or reports related to alcohol use, accidents or disruptions. You may be able to identify trends and areas of concern. If appropriate, speak to those concerned to gather their perspective on key issues. If you already use alcohol testing, analyse past test results not only from the perspective of identifying test fails but also to consider any challenges or difficulties in the testing process that could be resolved with policy changes. By assessing the effectiveness of existing policies, if any, you will more easily be able to identify gaps or areas for improvement.

Compare your organisation's alcohol policy and ethos against best practices in your industry and geographical region. A policy will have little effect if it is undermined by pro-alcohol workplace attitudes; if it is normal practice for managers to have lunchtime meetings in the pub or if alcohol is given as a bonus or reward for achievement, then any messages about the harm caused by alcohol will appear inconsistent.

In many ways, culture change is more difficult than policy change and it's harder to measure success, but the next few chapters will steer you through the opportunities and pitfalls that may arise. In the meantime, there are many actions that your organisation could instigate to support the aims of the policy, including health and wellbeing events, information and advice via the intranet, alcohol awareness training for staff, celebration events where alcohol is not served and more, that can help to tackle a boozy tradition.

5.3 STEP 2: POLICY DEVELOPMENT

> Employers must ensure they have a clear policy on substance misuse and we recommend that drug and alcohol misuse forms part of their wellbeing offering to ensure any need for disciplinary action is coupled with a preventative and supportive approach to help employees get the help they need.
>
> (Chartered Institute of Personnel Development, 2020)

Set Clear Objectives

Your objectives will determine your policy content and can also drive timescales. Ideally objectives should be SMART[2] and should outline clear benefits for the organisation and staff.

Remember though, your policy only one route to tackling alcohol-related harm. You should consider what other measures you will implement to complement your policy aims.

5.3.1 Baseline Data

At this stage, it's good practice to collect qualitative data on key metrics such as attendance, productivity and performance and incident reports. These data can be used at the evaluation stage to establish the effectiveness of your policy. Consider what you will need to record, how and why. It is unlikely that you will need to gather data on a particular employee, but should that be the case, make sure that you adhere to **GDPR** regulations.

5.3.2 Importance of Conducting a Risk Assessment

Assessing risk and evaluating the potential impact of alcohol-related incidents on safety and productivity is imperative. A risk assessment helps identify vulnerabilities within the organisation that may be exacerbated by alcohol-related incidents. This includes safety hazards, safety-critical job roles and areas where productivity may be compromised. By understanding the specific risks associated with your workplace, you can tailor prevention strategies to target the most pressing areas. A robust risk assessment provides the foundation for developing targeted, proactive prevention initiatives and it will also help to ensure compliance with safety legislation.

5.3.3 Involve Employees

For your policy to be a successful dynamic document, it makes sense to incorporate the views and ideas of those it affects. Involving employees in the policy development may be a resource intensive process, but it is certainly one worth considering. According to the Health and Safety Executive, it is essential to consult with staff, not only to inform them of the policy but also to seek and consider their views. There are a number of ways to achieve this—through focus groups, surveys or feedback sessions. Their insights are invaluable in shaping a policy that addresses their needs and concerns.

Sample Employee Survey Questions

The following set of questions is not intended as a definitive, but as a rough template for the type of questions that you may wish to ask. Consider what your aims are in asking these questions: do you want to ascertain whether there are any problems, discover what your employees know about alcohol, find out what people think of your existing policy? You may wish to consult with a working group on the set of questions that you will publish. Consider also, what format your survey will take—online, via interview, paper-based and how it will be distributed.

These scaled questions will give you quantitative data that will be simple to interpret but misses the nuances that can be gleaned with multiple choice or free text answers. For example, some of the questions mentioned earlier could be phrased as follows:

Do you ever have any concerns about a colleague's drinking?

No; yes occasionally; yes, often; yes, about more than one colleague

Or

What concerns do you have about the impact of alcohol consumption on your colleagues?

If you had concerns about your drinking, what sources of support would you consult?

Friends or family; GP; self-help materials; EAP or other employer-based support; local alcohol recovery service; Alcoholics Anonymous; sobriety coach; other (please specify)

Or

Where would you go for support if you had concerns about your drinking?

Which of these impacts does alcohol have on your workplace?

It fosters team bonding; increases absenteeism; it poses a health and safety risk; it eases the stresses of work; no effect

Or

How do you think alcohol consumption affects your workplace?

TABLE 5.1
Sample Survey Questions to Ascertain Employees Thoughts on Alcohol Issues in the Workplace

	Strongly disagree	disagree	Neutral	Agree	Strongly agree
Colleagues' alcohol consumption					
I have concerns about a colleague's drinking					
I have had to take on additional tasks or cover for a colleague who was impaired by alcohol					
One or more colleagues have been absent from work because of their drinking					
I have noticed reduced capacity, productivity or effectiveness in a colleague due to alcohol					
Own alcohol consumption					
I have had concerns about my own alcohol consumption					
If I was struggling with my alcohol use, I would feel confident in seeking support from my employer					
I have been absent from work due to alcohol					
I would know where to get help if I was concerned about my drinking					
I have drunk alcohol to reduce workplace stress					
I have a loved one whose drinking worries me					
Impact of alcohol on the workplace					
Drinking alcohol at work-related events enhances colleague relationships/bonding experiences					
Alcohol consumption is a health and safety risk in my workplace					
Alcohol consumption during the working day negatively affects productivity and performance					
Alcohol consumption can affect staff wellbeing					
Alcohol awareness					
I know the national guidelines on low-risk drinking					
I know how many units there are in a large glass of wine or a pint of standard strength beer					
I know and understand the legislation around drinking and driving					
I am aware of the potential mental and physical health implications of excessive drinking					

Alcohol policy awareness

My organisation has an alcohol policy

I am aware of the contents of my organisation's alcohol policy

I know how to access my organisation's alcohol policy

My organisation's alcohol policy offers a sensible response to alcohol consumption

If you're looking for qualitative responses, your approach could include one or more focus groups to gather contextualised opinions, stories and ideas. Focus group themes could include establishing what is working well about an existing policy and what needs improvement; what training should be available to managers or understanding the current impact of alcohol in the workplace. This is an opportunity for employees to share valuable perspectives, helping to highlight challenges that management may not be aware of. Feedback from staff allows your new policy to be bespoke to the organisation's unique needs.

Consultation not only explores the needs and wishes of your workforce but also encourages buy-in. If you can involve staff in the policy-making process you may be able to develop a more robust policy and collaboration from your workforce can minimise any resistance. It may also provide opportunities to open up honest, transparent conversations around alcohol harm at work. You will find additional wellbeing-focused cultural audit questions in Chapter 9.

CASE STUDY

A large multinational refreshed its alcohol policy to reflect its support offered to employees who identify that they have alcohol problems. In response to the successful recovery of an employee who had been supported through the recovery process, management were keen to promote the organisation's offer of help. Alongside a stringent drug and alcohol testing regime, in line with industry norms, the organisation was deeply committed to supporting employees.

A managers' focus group highlighted concerns, including the title of the new policy, which focused on 'testing management' and, at nearly twenty pages long, how inaccessible the document was to both employees and managers.

This prompted further discussion and solutions were found, including a new title and a one-page summary document that was easy to read and digest. Managers expressed increased satisfaction with the policy and praised the common-sense language of the new document.

5.3.4 Devise Your Policy Content

There are plenty of policy templates available online and lots of guidance about what to include. However, do beware of 'cut and paste' policies. Whilst it's useful to adopt a template as a basis for your own policy, I want to stress the importance of tailoring your policy to your organisation. In the event of an alcohol-related incident, you will rely on this document to protect both the organisation and employees. Livelihoods, and potentially lives, depend on getting this right. It's good practice to consult with industry and legal experts to ensure that your policy is meeting its objectives and there are no unforeseen loopholes. In litigation, a poorly worded (or poorly interpreted) policy can have huge implications for tribunal results.

The following sections suggest some of the areas you may wish to address in your policy. ACAS and the Health and Safety Executive have produced guidance on policy content, and both are a good source of information when developing your policy document.

5.3.5 Introduction

This should outline the rationale and aims of your policy. Usually, this will include a statement about the organisation's commitment to ensuring that the use of substances by an employee does not impair the safe and efficient running of the business or put at risk the health, safety or welfare of employees, customers, suppliers or members of the general public.

State the organisation's stance on alcohol use in the workplace, emphasising that safety, productivity and a harmonious work environment are top priorities. Express the commitment to providing support for employees facing alcohol-related challenges.

It is good practice to highlight the consequences of impaired job performance due to alcohol use, such as missed deadlines, subpar work quality or safety hazards. This is also an ideal place to state the organisation's responsibilities regarding relevant legislation.

5.3.6 Scope

It's useful at this stage to consider the scope of your policy. The key questions here are who, where and when? You should also establish whether there are any exceptions.

Who?	Does your policy apply to site visitors, including contractors, clients and suppliers? Does it apply to agency, casual and seconded employees, where their original employer may have different policies and practices? Does the policy apply equally to all staff, partners and directors? Are there additional considerations for safety-critical roles?
Where?	Does your policy consider off-site working? Are the rules applicable at all sites and work premises (including homeworking)? Are there clear rules around the provision of corporate hospitality? Does the policy cover employees' actions at work events, celebrations and other social functions?
When?	Are there limitations to drinking in the hours prior to working, for example in relation to safety-critical roles? Is there guidance on drinking and driving? Are there exceptions, during celebrations for example, or when sanctioned by the senior leadership team? Is provision made for employees returning home after events or corporate hospitality obligations? Is your policy applicable at work-related social events and outside of traditional working hours?

Contractors

Contractors are typically considered responsible for their own health and safety while on the premises of the employer, including managing their alcohol consumption. Employers may still have a duty to check that contractors' work does not endanger anyone, and you should incorporate alcohol-related policies and expectations into contracts or agreements. It's good practice to communicate your alcohol policy to contractors and require compliance when on-site. This could take the form of posters on-site or short written summaries to be handed to all contractors. Where possible, contractors should sign to acknowledge that they have been informed of your policy.

Part-Time Employees

Part-time employees should be subject to the same alcohol policies and expectations as full-time employees. You must ensure that part-time employees are aware of and understand the alcohol policy, regardless of their work hours; this includes staff on zero-hours contracts.

Visitors

Visitors, including clients, customers or guests, are typically not subject to the same alcohol policies as employees. However, employers should ensure that their premises remain safe, and that visitors' alcohol consumption does not endanger themselves or others. You may want to establish guidelines for visitors' behaviour, especially in environments where safety is a concern or where intoxication is an accepted risk, such as on licenced premises.

5.3.7 Definitions

Define key terms, such as 'intoxication and 'dependence' to avoid ambiguity so that everyone understands the policy's principles and its practical application. Vague terms, such as 'under the influence', can be interpreted in different ways (e.g. intoxicated, any positive blood or breath screening, testing as above the drink-drive limit, impaired in performance through alcohol). It's also useful to include a glossary of terms. The glossary at the end of this book will provide some ideas and definitions that you can use.

5.3.8 Roles and Responsibilities

All employees and managers need to be aware of their responsibilities regarding the policy. For example, managers will need guidance on how to determine that policies and practices are adhered to, and how to identify and respond to problems promptly. Employees should understand their obligation to arrive and remain fit for work, unimpaired by alcohol. Rules around alcohol consumption before and during work activities should be stated explicitly and any exceptions clearly given. Your policy may also require employees to disclose problematic alcohol use or seek medical advice or report colleagues who are in breach of the policy.

Your policy will need to state whether suspicion of a breach may lead to alcohol testing, searches or other measures to confirm an infringement. The consequences of refusing a test or search should also be clearly stated.

5.3.9 Prohibited Conduct

Specify when and where alcohol consumption is permitted or prohibited, including during work hours, at company events and on company premises. Prohibited activities may include possession or distribution of alcohol on company property and alcohol-related harassment or discriminatory behaviour, such as sober shaming.

If relevant, explain the organisation's stance on alcohol testing and searches, including under what circumstances testing may be conducted, for example in response to reasonable suspicion.

You should clearly outline the expectations for safe driving within contracts and policies. This includes restrictions on alcohol consumption before and during work hours, especially for roles that require an employee to drive, whether in a company vehicle or not.

5.3.10 Testing and Screening

Outline any alcohol testing or screening procedures and the circumstances under which they may be conducted. Take some time to reflect on the purpose of your testing regime and consider how best to express this. Further information about alcohol testing can be found in Chapter 7.

In some sectors, pre-employment drug and alcohol testing is common. Testing will take place with the same rigour as for employees and potential candidates must be informed in advance of any testing requirements. It's good practice to include this information in your policy.

5.3.11 Consequences

Clearly state the consequences of violating the policy, including disciplinary measures that may be applied. State any legal requirements for your industry. It's a good idea to align breach procedures with your disciplinary policy. As a supportive employer, it will be important to differentiate between disciplinary measures in response to a breach and support for an employee disclosing an alcohol problem and requesting help. Where an employee chooses to disclose their difficulties only after having been caught in breach of the rules, careful consideration needs to be given to whether a conduct (disciplinary) or capability (supportive) route will be followed. See Chapter 4 for further guidance.

You can implement a range of disciplinary measures when an employee is found to be impaired by alcohol. These may include verbal warnings, written warnings, mandatory counselling, suspension or dismissal. Be clear about escalating sanctions for repeated infringements if applicable and take steps to ensure that your policy sets out measures that are fair and consistent.

CASE STUDY

An employment tribunal in Dundee found in favour of a council employee who became intoxicated at work. His dismissal was determined to be unfair as the council failed to adhere to their policy, which stated that employees with addiction issues would be given an opportunity to seek help and not purely subjected to disciplinary action. Previous incidents of alcohol abuse by the claimant had not been investigated by the employer but demonstrated an ongoing problem that should have triggered an offer of support.

(Angus Council v Edgley, 1999)

5.3.12 Managing Customer and Client Intoxication

If your organisation has roles that are customer-facing or where employees interact with the public, they may at some point need to how to manage situations where customers or clients exhibit intoxication. If the likelihood of this kind of interaction is low, it can be addressed in your policy, although you may have other policies that deal with this. Where this may be a frequent occurrence, such as in the service or hospitality industry or in retail, training and support for staff are essential. Dealing with individuals who are impaired through alcohol can be frightening and confusing and a situation can quickly escalate. Training needs to be ongoing, and you should incorporate any lessons learned from serious incidents. Role-playing scenarios, though rarely popular, can be tremendously helpful in giving employees the confidence and skills to respond effectively to intoxication in customers. You will find further guidance in Chapter 6.

5.3.13 Employee Assistance

Employees who recognise that they have an alcohol problem should be encouraged to seek help, in confidence, at an early stage. Bear in mind, however, that alcohol problems are notoriously difficult to identify. As alcohol is legal and freely available, it can be hard for someone to recognise that their drinking is having a negative impact. Indeed, different people can have vastly different outcomes from drinking similar amounts of alcohol in both the short and long term. To help to overcome this, alcohol education needs to form part of your wider wellbeing offer. More about this in the coming chapters.

Describe the in-house resources available to employees for seeking help, including EAPs or counselling services. You may also wish to include:

- A list of local and national services to which employees can self-refer
- Advice to consult with their GP
- A list of services for anyone affected by a loved one's drinking
- National helplines
- Details of crisis support services
- Information about mental health support
- Links to online screening tools, such as the AUDIT

5.3.14 Confidentiality

Address the confidentiality of information relating to disclosures of alcohol problems and outline your compliance with relevant **GDPR** rules. Where possible, seek the employee's permission to disclose to relevant parties, such as their line manager.

Even without permission, if a safeguarding issue is identified, it may be necessary to inform others of details concerning the employee's fitness to work.

Outline the procedure for employees to raise concerns related to alcohol, whether about their own issues or those of their coworkers. Guarantee anonymity and confidentiality for individuals who report these concerns.

5.3.15 Review and Approve

This is another excellent opportunity to involve your workforce. You could set up a focus group or distribute your draft document for consultation within teams. Input from employees, managers and HR promotes transparency so that different perspectives are considered. This may be an iterative process as new ideas often arise at this stage.

Submit the policy draft for review and approval by legal counsel and relevant stakeholders. Besides ensuring legal compliance, it's useful to sense-check your policy for clarity and consistency. You can then make any necessary revisions based on feedback. A well-reviewed policy helps to safeguard your organisation's reputation by demonstrating internally (and externally) that it takes these issues seriously.

5.4 STEP 3: DISSEMINATION

Develop a communication plan for rolling out the policy, including how it will be presented to employees, when it will take effect and how questions or concerns will be addressed. Effective communication is key to ensuring employees understand and comply with the alcohol policy. According to a recent CIPD survey, the top three methods of communicating drug and alcohol policies are: via the staff handbook, as part of the onboarding process and via the intranet.[3]

5.4.1 Acknowledgement

You will need to make it clear to employees that they are required to read and acknowledge their understanding and acceptance of the policy. A simple signature to an acknowledgement document should suffice to evidence that the policy has been disseminated to all staff and that they are then fully informed of their responsibilities and any support provision therein.

5.4.2 Workshops and Training

Conduct workshops, webinars or training sessions to educate employees about the policy's content, objectives and implications. This can be coupled with alcohol awareness events as part of your wellbeing initiative. It makes sense to link the launch of your policy to your wellbeing calendar or an organisational milestone.

I'm often surprised at how little knowledge most people have about alcohol. It's ubiquitous, most of us drink it, society has a complex relationship with it. But when it comes to understanding units, safe limits, risks, intoxication and dependence, many, many myths and misapprehensions abound. Preparing your workforce in this way also anticipates any problems later, and a good alcohol education is a must for a healthy workforce. The charity Alcohol Change UK has developed a framing toolkit for talking about alcohol, that is available on their website. It advises on how to share effective alcohol harm messages. It's essential to avoid scaremongering or shock tactics, focusing instead on the range of wellbeing impacts, the power of pro-alcohol messages and steps that anyone can take to reduce risks, without marring their enjoyment.

5.4.3 Manager Training

Whilst the majority of organisations appreciate the need for an alcohol policy, and around 75% currently have a policy in place, only around one in eight of these train line managers in how to interpret and implement the policy. Just one in three provide guidelines on how to respond to disclosure of an alcohol problem or how to signpost staff to support.[4]

Before policy launch, you'll need to ensure that your managers are fully apprised of their role, particularly if they will be involved in responding to alcohol-related incidents or disclosures or when implementing an alcohol testing regimen. Often, although managers already possess the skills needed to carry out these duties effectively, they may lack confidence in talking to employees about drinking. This is a common fear, and one that good skills-based training can overcome.

Ideally, manager training should be mandatory so that anyone with direct reports has skills and knowledge in the following areas:

- Ethos and content of the policy
- How to interpret the policy in practice
- Screening employees for alcohol issues and delivering brief advice
- Responding to disclosures
- Talking to an employee about their alcohol use
- Issues of confidentiality
- Your testing programme and their role in its implementation
- Supporting colleagues who have identified an alcohol problem
- Signposting and referral—what help is available?

Although line managers are likely to play a key role in implementing the policy, responding to problems should not solely be the responsibility of management. It should form part of the ethos of the company and be embedded in procedures and practice as well. Others who may be involved in this include OH services, HR, the leadership team, employees, unions and customers. An effective alcohol policy will also dovetail with other organisational policies to provide a consistent message:

- Health and Safety
- Recruitment
- Lone Working
- Disciplinary
- Data Protection
- Equal Opportunities
- Sickness Absence
- Risk Management

5.4.4 Employee Awareness

Provide clear mechanisms for employees to seek clarification, report concerns or request support related to alcohol issues. Be clear about what anonymity can be offered. Transparency about how the new policy might impact staff, what support is available and how to access help and support will increase compliance when the policy is rolled out.

It's good practice to implement a 'grace' period—usually three or six months—between the launch of the policy and its implementation. This allows employees who may be having difficulties with their drinking some time to reflect and consider seeking help or making efforts to change their drinking behaviours before the policy comes into effect. During this period, you can implement employee or manager training, produce education materials and alcohol-focused wellbeing initiatives can be offered.

There are many innovative ways of introducing your policy. I've worked with companies that have offered free self-breathalysers to employees during the grace period. Others have offered free online coaching to anyone who wants to explore their drinking habits. One organisation set up a dedicated hotline and email address for employees to seek guidance or report concerns anonymously. Yet another invited a former dependent drinker in to talk about their experiences of recovery. Whatever you choose, a grace period is a show of goodwill that benefits both the organisation (allowing an opportunity to tackle unforeseen problems) and employees (who have time to familiarise themselves with any changes).

Create user-friendly materials, such as pamphlets or infographics, summarising the policy's key points for easy reference. You could use a simple dos and don'ts list or a bullet-pointed summary; incorporate graphics or put on an interactive webinar. Simplicity and clarity are key here.

You may wish to incorporate alcohol-related considerations into your recruitment process, too. You can include questions in interviews or application forms that address an applicant's willingness to comply with the alcohol policy. This may be particularly relevant in sectors such as hospitality, where working in an alcohol-fuelled environment may be a draw for candidates. Additionally, it's useful to educate recruiters and hiring managers about the policy's provisions to ensure they communicate expectations clearly to candidates.

Make sure your alcohol policy and associated practices do not unfairly target employees or job applicants with alcohol-related issues. As required by law, offer reasonable accommodations to employees seeking assistance for alcohol-related challenges. Additionally, it is advisable to avoid basing hiring, promotion or disciplinary decisions solely on a person's history of alcohol misuse.

5.5 STEP 4: MONITORING AND EVALUATION

If you collected quantitative data prior to the development of your policy, you could now implement a programme of evaluation to monitor how effective your policy is against your key metrics. By analysing data on alcohol-related incidents, policy violations and absenteeism, you'll be able to assess how well your policy is fulfilling its aims. Of course, qualitative data, such as employee satisfaction, should also be gathered. Regular monitoring is important so that you will be able to identify any difficulties at an early stage.

Conduct periodic reviews of the policy to assess its relevance and make updates as needed. This will generally be every two to three years, or when there are major organisational or sector changes that will impact the policy. You can also ensure ongoing training and education for employees to keep them informed about the policy's provisions and any changes.

Remember, though, that your policy is just one strand of your alcohol management strategy. If you don't back it up with appropriate boundaries and action, and if you don't support employees who are finding it difficult to manager their drinking, your policy will not be effective. You will find further information on data collection and monitoring in Chapter 10.

NOTES

1 CIPD, 2020. Managing Drug and Alcohol Misuse at Work (https://www.cipd.org/globalassets/media/knowledge/knowledge-hub/reports/drug-alcohol-misuse-work-report-1_tcm18-83090.pdf).
2 Specific, Measurable, Achievable, Relevant and Time-bound.
3 CIPD, Managing drug and alcohol misuse at work, 2020 (https://www.cipd.org/globalassets/media/knowledge/knowledge-hub/reports/drug-alcohol-misuse-work-report-1_tcm18-83090.pdf).
4 CIPD, Managing drug and alcohol misuse at work, 2020 (https://www.cipd.org/globalassets/media/knowledge/knowledge-hub/reports/drug-alcohol-misuse-work-report-1_tcm18-83090.pdf)

6 Managing Alcohol-Related Incidents

6.1 INTRODUCTION

Under UK law, employers have a duty of care to ensure the health and safety of their employees. This includes taking reasonable steps to prevent employees from working under the influence of alcohol, responding to reports or suspicion or reports of intoxication and, where appropriate, providing support and assistance to employees facing alcohol-related issues.

Prevention is better than cure, certainly on the topic of alcohol-related incidents, but no matter how carefully prepared your policy and how supportive you are of your workforce, incidents and accidents remain a risk. In this chapter, we look at best practices for responding to various alcohol-related incidents, such as intoxication, accidents and disciplinary issues. We will also consider drink-driving offences and how these may be grounds for dismissal, or not.

QUESTIONS FOR REFLECTION

Thinking about your workplace, what immediate risks can you identify from:

a. An employee being hungover at work
b. An employee driving whilst under the influence of alcohol
c. An intoxicated client or customer

Consider morale, reputational and financial risk besides safety and the implications for productivity.

6.2 RESPONDING TO ALCOHOL-RELATED INCIDENTS AND INTOXICATION AT WORK

6.2.1 Intoxication in the Workplace

When an employee is identified as or suspected of intoxication at work, you need to act swiftly and in accordance with your established procedures. There are generally four routes by which an individual may present as alcohol-impaired at work:

- Working in a debilitated state post-intoxication (e.g. through a hangover)
- Arriving at work in a state of intoxication

DOI: 10.1201/9781032648927-7

- Drinking at work in the course of their duties, including whilst providing hospitality for clients
- Drinking during breaks, lunch or whilst otherwise being temporarily 'off the clock'

Whilst none of these is indicative of an alcohol addiction per se, it is likely that an employee who finds it hard to refrain from drinking before or during work may generally have difficulties controlling their alcohol consumption. Of course, there may be times where unusual circumstances—overdoing it at the Christmas party, for example—may play a role, but it is often found that a pattern of such behaviours comes to light once the person is challenged about their alcohol consumption. Whatever the underlying reason, it is always cause for concern when someone is demonstrably intoxicated. Your policy (see Chapter 5) should give guidance on how to proceed, but there are key steps that need to be followed to maintain the safety of the individual and others.

As it is rarely the case that an employee freely admits to intoxication or shows such overt behaviours that there can be no doubt, it is wise to proceed under the guise of suspicion of intoxication. There are several illnesses that can mimic drunkenness so consider whether this may be the cause of the behaviour. Signs of intoxication may come to your attention without a precipitating incident, but it may also be because of an accident, error, incident or near miss. Whether intoxicated or ill, your employee is vulnerable and will need help and support at this stage. The immediate response will be comparable in either case, and the safety of everyone involved is paramount.

MENTAL CAPACITY IN RELATION TO ALCOHOL INTOXICATION

According to the Mental Capacity Act:[1]

> *For the purposes of this Act, a person lacks capacity in relation to a matter if at the material time he is unable to make a decision for himself in relation to the matter because of an impairment of, or a disturbance in the functioning of, the mind or brain.*

It is therefore important that when an employee is intoxicated in the workplace, the employer considers whether they are able to make sound decisions and act in their own best interests.

As a precaution, ensure the employee's immediate safety by removing them from the workplace, arranging transportation home and later engaging in a supportive discussion with the employee when they are sober.

6.2.2 Procedures for Reporting Alcohol-Related Incidents

It's good practice to establish clear, unequivocal protocols for reporting incidents. Your policy should guarantee the confidentiality of those reporting and encourage an environment of accountability among employees. Outline how and to whom it is best to report and offer reassurances that prompt, accurate reporting is necessary for health and safety.

Actively encourage coworkers who observe signs of alcohol-related incidents in their colleagues to share their concerns. They should document their observations, including the date, time, location and specific behaviours witnessed. Let them know that their observations can help create a safer work environment and that they are not helping colleagues by overlooking potential hazards. Support accountability by emphasising that reporting concerns is a responsible and supportive action. Your policy and ethos should also communicate the organisation's commitment to addressing alcohol-related incidents discreetly and professionally.

Nevertheless, it can be challenging for an employee to report a colleague's behaviours so you may wish to promote the use of anonymous reporting options to alleviate concerns about potential retaliation. In larger organisations, this could be a designated manager or HR representative or a secure online reporting portal. For smaller organisations, it may be a line manager or wellbeing champion. If your organisation has an incident reporting form, employees should complete it, providing as much detail as possible about what they observed. This form should be designed to protect the identity of the reporter. Assure coworkers that their identities will be protected and acknowledge and thank employees who report concerns. Check that employees are aware of their rights in this regard.

When a supervisor or manager receives a report of an alcohol-related incident, they should take it seriously and initiate a response immediately. Stress the importance of maintaining strict confidentiality during the investigation process. Sensitive information should only be disclosed to those directly involved in the incident response. Supervisors or managers should document all incident details, including the nature of the incident, names of individuals involved, and any actions taken. This documentation should be stored securely. If the incident poses an immediate safety risk, the supervisor or manager should take immediate action to guarantee the safety of employees and address any potential hazards.

6.2.3 Accidents Involving Alcohol Impairment

When workplace accidents occur, it is not often that alcohol is immediately considered a causal factor, yet up to 40% of accidents at work are related to alcohol use.[2] It's important to consider the role that alcohol may play, whether through an employee suffering the after-effects of a heavy drinking session, or where there is evidence of a longer-term problem.

Knowingly allowing an intoxicated employee to continue working, where there is risk to that employee or others, can lead to prosecution,[3] so all employees need to be clear about their responsibility to ensure that they are in a fit state to work and that they must not 'cover' for colleagues who may be intoxicated. Your policy and other measures will help in this respect, but as they say, accidents do happen, and when they do, it's wise to consider any evidence that drinking may have been a contributory factor. Your health and safety policy or accident policy will cover practical responses to workplace accidents, but outlined next, you will find some additional guidance on how to proceed if alcohol impairment is suspected as a factor.

6.2.4 Immediately Post-Incident

Address the immediate safety and medical needs of those involved in the accident. Check whether anyone, including the intoxicated employee, is injured and provide first aid immediately. Look for visible injuries such as bleeding, bruises or fractures and assess whether emergency medical services need to be called. Any head injuries or signs of severe intoxication (e.g. vomiting, confusion or stupor) should prompt immediate notification of the emergency services. Although alcohol misuse may be responsible for some symptoms, the employee may also be injured. Alcohol can act as an anaesthetic, so an intoxicated person may not feel pain from injuries immediately. Alcohol also dilates blood vessels and can reduce clotting factors so any bleeding may be more severe. Call for medical personnel or first aid so that anyone affected can be treated or stabilised.

Clear the area around the incident to prevent additional hazards, and to ensure the area is safe for responders to access. Take steps to prevent other employees from entering if there's any potential danger, by using hazard tape, cones or signs to block off the area. This also preserves any evidence related to the accident's cause. Take clear photos or videos of the area and capture any relevant details, including machinery, tools or objects involved in the incident, showing their settings and state (e.g. on/off position, damaged or in the wrong place).

Document environmental conditions, such as lighting, floor conditions (e.g. wet or dry), temperature, ventilation and any visible hazards like spills or obstacles. If possible, take note of where the employee was when the incident occurred. Later, someone under the influence of alcohol may remember events differently, so careful documentation of the incident scene can provide requisite information about how the accident unfolded and is vital for investigation and future accountability.

Conduct an initial assessment to determine if alcohol impairment may have contributed to the incident, taking into consideration any observable signs of intoxication. If your organisation's alcohol policy allows for post-accident testing, request an immediate test. Alcohol testing should be conducted promptly as alcohol levels decrease over time. Testing methods may include breathalyser tests or blood samples, depending on your policy and the nature of the employee's role. Request that the employee consents to the test and maintain records of consent forms and test results confidentially. This documentation is important for legal protection and ensures compliance with privacy regulations. You'll find guidance on workplace testing in Chapter 7.

Isolate and secure any physical evidence of alcohol consumption. If bottles, cans or other containers are found in the vicinity, take photos and secure them in a safe location. Avoid searching the employee's personal belongings unless expressly allowed by company guidelines and applicable laws or where the employee freely gives written consent. Keep the scene intact for investigation and do not disturb any evidence that could suggest alcohol use until it has been thoroughly documented and photographed. This will improve evidence preservation and may prevent legal challenges later.

6.2.5 RIDDOR

In the event of a death or injury in the workplace, you may need to make a report under the Reporting of Injuries, Diseases and Dangerous Occurrences Regulations (RIDDOR). Guidelines on the requirements for RIDDOR and how to report can be found at hse.gov.uk. Where alcohol may have contributed to the injury or death, this must be included in the report.

6.2.6 Company Events and Corporate Hospitality

Naturally, it is advisable to ensure that company events do not rely solely on alcohol and to offer non-alcoholic alternatives (see Chapter 9 for alcohol-sensitive event planning) but work events, parties, celebrations and corporate hospitality are social in nature and may well involve opportunities to consume alcohol. In essence, responding to alcohol-related incidents that occur at these events, whether on or off-site, during or outside of office hours, requires a comparable response to any other concerning alcohol-related occurrence. Your alcohol policy should state guidelines for the provision and consumption of alcohol in all work-related circumstances. However, it's also important to establish beforehand what the rules around alcohol consumption are and conduct a thorough risk assessment. Potential risks include unacceptable behaviour, such as overconsumption of alcohol, harassment, sexual assault, discrimination, violence or damage to property.

The assessment should consider factors such as venue safety, alcohol service, transportation arrangements and potential behavioural risks associated with drinking. Special attention should be given to safeguarding vulnerable individuals and ensuring there is no risk of employees engaging in dangerous activities, such as driving under the influence.

Practical measures to manage alcohol consumption at events are also good practice and employers should limit the availability of alcohol to discourage overindulgence. This might include setting a cap on the number of free drinks provided or using a voucher system. Non-alcoholic beverages should be prominently available and appealing, facilitating an environment where employees do not feel pressured to drink. To avoid promoting excessive drinking, activities like open bars, drinking games or contests involving alcohol should be avoided. You might also consider offering food throughout the event, which can help moderate the effects of alcohol consumption.

Designating responsible staff members to oversee the event and act as points of contact in case of issues is another best practice. The designated person, often a manager or other senior employee, should remain sober to handle any situations that arise. Their responsibilities might include monitoring behaviour, addressing any immediate concerns and providing support to employees who may be distressed or unwell. They should be trained to handle incidents calmly and impartially, intervening early if signs of potential problems, such as escalating conflict or intoxication, become apparent.

If an alcohol-related incident occurs, you must, as always, respond quickly and appropriately. Begin by assessing the situation to determine what happened and separating those involved if necessary to de-escalate tension. Your response should focus on safety, with medical assistance sought if required. Any investigations should follow established disciplinary procedures, ensuring a fair and transparent process. Employers should be prepared to involve external authorities if the incident involves potential criminal activity, such as assault or drink-driving.

Follow-up actions after an incident are equally important. Set up a review of events to identify any weaknesses in the planning or execution of the event that may have contributed to the issue. This reflection can inform improvements for future events, such as stricter alcohol controls or enhanced training for staff. Providing support to employees involved in or affected by the incident may also be necessary. This could include access to counselling services, an EAP or mediation to repair workplace relationships.

Finally, inclusivity should be a priority in planning workplace events. Non-alcoholic options and engaging activities that do not revolve around drinking should be integral to the event's design, ensuring that employees who do not consume alcohol, for religious, health, personal or any other reasons, feel equally included.

CASE STUDY

MBNA held a 20th-anniversary celebratory work event at a racecourse where normal workplace conduct standards were expected. During the event, Mr Jones and Mr Battersby engaged in behaviour that escalated into an altercation. This began with seemingly light-hearted banter, including Mr Battersby kneeing Mr Jones in the leg and Mr Jones licking Mr Battersby's face. However, it escalated later when Mr Battersby kneed Mr Jones again, prompting Mr Jones to punch him in the face. After the event, Mr Battersby sent aggressive and threatening messages to Mr Jones.

Both were disciplined, with Mr Jones being dismissed for gross misconduct, largely due to the punch and his behaviour at the MBNA branded event, which MBNA argued could damage its reputation. Mr Battersby, who had sent threatening texts, received only a final written warning. Mr Jones claimed unfair dismissal, arguing that the sanctions were inconsistently applied.

The Employment Tribunal initially found the dismissal unfair, suggesting that MBNA's inconsistent sanctions undermined the fairness of the process. However, MBNA appealed successfully. The Employment Appeal Tribunal ruled in MBNA's favour, stating that the Tribunal had failed to properly assess whether the situations of Mr Jones and Mr Battersby were truly comparable. The Appeal Tribunal emphasised that differences in individual circumstances can justify different disciplinary outcomes, particularly since Mr Jones initiated physical violence during the event.

IMPLICATIONS FOR EMPLOYERS

This case underscores the importance of treating similar cases fairly but also highlights that disparate disciplinary actions are permissible if there are valid, case-specific reasons. Employers must:

Conduct thorough investigations and make sure the facts of each individual's behaviour are considered.
Document clear reasoning for applying different sanctions to employees involved in the same incident.
Be consistent in applying policies but understand that fairness does not always mean identical outcomes.

The case establishes that inconsistent sanctions alone do not automatically make a dismissal unfair. Instead, fairness depends on whether the employer's decisions are reasonable and well-founded, considering the circumstances. Employers are afforded discretion in their responses, provided they act within the boundaries of reasonableness.

(*MBNA v Jones*, 2015, UKEAT/0120/15/MC)

6.3 POST-INCIDENT RESPONSE

6.3.1 Ensure Immediate Safety

Under the Health and Safety at Work Act 1974, employers have a legal duty to ensure the safety of staff, so you'll need to quickly evaluate the severity of the situation. If the employee is showing signs of slurred speech, unsteady gait or erratic behaviour, assess whether they pose any immediate physical danger to themselves or others or any risks to company property or reputation. Consider factors such as their role, their behaviour and their location. If their role involves high-risk tasks, such as driving or operating machinery or if they could cause harm due to impaired judgement, coordination or reaction times, immediately isolate the employee from these tasks. Politely but firmly request they step away, citing safety as the priority. If there's any sign of the employee becoming aggressive or non-compliant, have security or a senior manager on standby. Safety is the primary concern, so this step may be essential in certain workplaces, such as warehouses or healthcare facilities, where physical safety risks could be extensive.

Ask the employee to accompany you to a private, quiet location to maintain confidentiality and avoid distractions whilst you investigate further, ideally choose an HR office or meeting room. This should be done tactfully and as privately as possible. Respect the employee's dignity and address the issue as discreetly as pas you can. Drawing attention could not only harm their reputation but also escalate the situation or cause distress among other staff members.

For accountability and safety have a neutral observer, such as a manager, HR representative or designated witness present during any interactions. A third party can also help to defuse any potential conflict and provide backup should the situation worsen or if there is any later dispute about how the incident was handled.

Allow time for honest responses; giving the employee space to explain or admit to drinking (if that is the case) can ease tension and provide context for your next steps. If the employee appears unwell but does not admit to drinking, acknowledge that medical or other personal issues could be at play. Stick to what you objectively observe rather than assuming the cause.

6.3.2 Is It Alcohol or Something Else?

Observe for signs of intoxication and the smell of alcohol. You can find a list of common signs in Chapter 4, but these do not confirm intoxication without careful assessment. Ask respectful observation-based questions to clarify the person's state. Avoid jumping to conclusions, but ask questions that allow them to share their condition: 'You seem to be having trouble balancing; is everything alright?' or 'I noticed some changes in your speech today; have you been feeling okay?' as this can help to determine whether another factor is involved. You may already be aware of their medical history and any conditions that could present as signs of intoxication, or this may not be something that your employee has disclosed. Several acute and chronic mental and physical health conditions can cause confusion, tremors, slurred speech, stumbling, lethargy, erratic gait, seizures and out-of-character behaviours. Avoid immediate accusations to maintain a respectful and legally sound approach.

In most cases of visual intoxication, sending the employee home is appropriate. Clearly communicate that the employee is being sent home for their wellbeing and workplace safety. Let them know that a follow-up meeting will be scheduled to discuss the incident. This will be important for an investigation and prevent accusations of unfair or rushed action. Inform them that this action is standard protocol and not necessarily disciplinary. Emphasise that the goal is to gather information and handle the matter fairly. Allowing them to remain on-site could impact their ability to work safely and put the organisation at risk. Arrange for safe transportation home or arrange medical care as necessary. You may need to contact a next of kin to collect them, particularly if they seem in distress. Under no circumstances should an impaired employee be allowed to drive themselves home. If they insist, explain the legal and safety implications clearly and request their compliance for their own and others' safety.

If the employee's condition appears more severe—such as showing signs of confusion, vomiting or fainting—call for medical assistance. Record this step as part of your duty of care. If the employee becomes unconscious, get urgent medical help and place them in the recovery position.

6.3.3 Alcohol Testing

If your organisation operates a 'with cause' testing regime, a breath alcohol test should be administered at this time, by a qualified, competent person. If the employee

TABLE 6.1
Health Conditions That May Mimic Intoxication

Epilepsy	Brain injury
Diabetes	Fever
Stroke	Urinary tract infection
Multiple Sclerosis	Meningitis
Parkinson's Disease	Huntington's Disease
Dementia	Fatigue
Medications	Menopause

tests 'non-negative' for alcohol (or positive above the company-endorsed limit), you should proceed in line with your alcohol policy. If the test is negative, you can explore other potential causes, such as drug use, medication or a health condition. Guidance on testing procedures can be found in Chapter 7 and considerations about capability and conduct of an employee are discussed in Chapter 4. If your organisation employs an outside agency to administer tests, their advice should be sought immediately. There may also be union or trade association concerns or regulations, so seek guidance from the relevant bodies.

6.3.4 Next Steps

If an incident has occurred that might constitute a criminal offence, consider contacting the police. In the event of assault, drink-driving or wilful damage to property, this is likely to be the case. Again, your careful documentation of events may be required evidentially in court so be mindful that you are able to provide a thorough and contemporaneous account.

Ideally, the employee should remain under paid suspension whilst an investigation is conducted. This approach provides time to investigate thoroughly and ensures fair treatment. In the event of such incidents, your alcohol policy is invaluable. It should provide guidance on how to proceed and the employee can be reminded of its contents and their obligations as well as having access to the procedures that you will implement in response. Without a policy, there may be confusion about roles, timescales and the potential consequences arising from the incident.

If paid suspension is not possible, your follow-up needs to be conducted as swiftly as possible without compromising thoroughness. Whatever the outcome, this will be a stressful time for all concerned, especially the employee at the centre of the investigation. Keep them up to date with your progress and offer whatever help you can along the way, such as signposting to treatment options or their GP.

Document the details of the incident at your earliest opportunity. Stick to factual observations such as the time, location, who was present, the employee's behaviour, statements given and actions taken. It's important to take statements from any employees involved in the incident and other witnesses. This can include specific comments made by the employee under investigation and any direct actions they took. Check that all documentation is factual, objective and free from personal opinions and judgemental language. Comprehensive documentation is vital in the case of a formal investigation or if disciplinary action is necessary.

There may be some speculation among other employees about the incident and it is good practice to remind everyone of confidentiality and quash any rumours as they arise. Keep all details of the investigation and meeting confidential. Make sure that those involved in the investigation understand the need to maintain the employee's privacy.

6.3.5 Formal Investigation

Conduct a fair and objective investigation according to the organisation's disciplinary procedures and alcohol policy. If the company has an HR or legal team, involve them to establish legal compliance and objectivity. Timescales are important. You should proceed with haste to reduce any unnecessary worry on the part of the employee but ensure that you take the time to meet all procedural requirements.

Schedule a formal meeting to discuss the incident and any underlying issues that may have contributed to alcohol misuse. The employee should have the opportunity to invite a colleague, union rep or friend to attend. Approach the conversation professionally and allow the employee to explain their behaviour, whether it was an isolated incident, any extenuating circumstances or a sign of an ongoing issue. Be open to any reasonable explanations, such as medication side effects or stress-related issues, which could have contributed to their behaviour.

Review the employee's work history, any previous disciplinary issues and their general behaviour. Any disciplinary measures should reflect the severity and impact of the incident. For example, minor incidents with no harm may warrant a warning, while a serious or repeated offence may require more severe action. Apply disciplinary measures as consistently as possible so that the actions align with what's outlined in the alcohol policy and to ensure there is no perception of favouritism or unfair treatment. Consider less severe penalties, such as a warning or mandated coaching or counselling, for first-time or minor incidents. For repeated or high-risk incidents, disciplinary action up to and including dismissal may be necessary.

6.3.6 The Offer of Support

If the employee admits to alcohol-related problems or if it becomes apparent that they may benefit from support, refer them to an EAP or provide information about external counselling services. This not only shows empathy but also helps address the root causes of the behaviour. If the employee's situation does not call for immediate dismissal and they demonstrate a willingness to seek help, consider additional support, such as a leave of absence for rehabilitation, temporary reassignment to a lower-risk role or part-time working.

This can demonstrate the company's commitment to employee wellbeing while setting clear expectations for improvement.

If the employee is willing to seek help, work with them to create a plan for a structured return to work. This might include regular check-ins, reduced hours initially or reassignment to a less safety-sensitive role. A non-punitive approach, where the employee is encouraged to seek help without fear of immediate dismissal, can motivate them to address any underlying problems and re-engage positively with work.

Set parameters for the support offered. Alcohol dependence is a chronically relapsing condition, and you will need to consider the financial and business implications of supporting an employee in the longer term. If an employee is unable to carry out their normal duties for a substantial length of time due to alcohol dependence, they should be treated as any other employee with a long-term health condition and eventual termination of employment may be the result. However, many employees in this position *do* recover and continue to have long successful careers with their organisation.

QUESTIONS FOR REFLECTION

What support does your organisation currently offer to employees who have an AUD?
For how long would the organisation realistically be able to support an employee in recovery?
Who would be responsible for monitoring progress?
What would your criteria for successful rehabilitation be?

Consider the costs of support including changes in employee workload or attendance, management time, potential therapy costs etc.

6.3.7 Remote and Homeworking

Managing alcohol-related performance issues during homeworking can be more difficult. The challenges differ from onsite management due to the remote context, reduced visibility and the private nature of home-working environments.

In a home-working scenario, managers must rely on changes in behaviour, productivity and communication rather than direct observation. Employees who frequently miss virtual meetings, show inconsistent work quality or demonstrate unusual communication patterns (e.g. being withdrawn or erratic in tone) may be exhibiting signs of alcohol-related issues. But of course, these issues could be caused by a number of different factors. Regular one-on-one virtual check-ins can help managers identify concerns early. Whether the cause is alcohol or other difficulties that the employee is experiencing, managers should be confident to respond promptly. It is generally a lack of confidence, not skills, that causes managers to shy away from tackling alcohol-related issues. This can be ameliorated by having clear guidelines in place.

Your alcohol policy should explicitly address alcohol consumption during work hours, including remote work. The policy must be communicated clearly and reinforced through training and employee handbooks.

If performance issues persist, employers should document all incidents thoroughly. This includes missed deadlines, unresponsiveness and any other conduct or capability-related concerns. Documentation is important for ensuring fairness and protecting the company from potential legal challenges. If disciplinary action is necessary, virtual meetings should be conducted professionally, with HR present when appropriate. Employers should make use of secure platforms and communicate the purpose and format of the meeting in advance.

6.3.8 Differences When Managing Onsite Employees

For onsite employees, alcohol-related issues can often be identified more easily through direct observation. Signs such as slurred speech, the smell of alcohol or impaired motor skills are noticeable in person, allowing for immediate intervention. Managers can address concerns in real time, potentially preventing safety hazards. On the flipside, in onsite environments, alcohol misuse can pose safety risks that may be mitigated by home working. This is not to say that incidents and intoxication should not be treated with the same gravity as onsite problems, and employees should be made aware that policy breaches, wherever they occur, will incur consequences.

Regardless of the work setting, employers must handle alcohol-related issues in compliance with UK employment laws. This includes respecting employees' privacy rights, ensuring any disciplinary action is fair and proportionate and offering reasonable accommodations for those seeking treatment for substance misuse. Employers should also consider the ethical implications of monitoring and intervention. In remote settings, invasive surveillance methods are generally discouraged and instead, a focus on trust, open communication and support is recommended.

6.4 BEST PRACTICES FOR RESPONDING TO AN INTOXICATED CUSTOMER

Over half of UK adults consume alcohol at least once a week,[4] and over a quarter of these admit to binge drinking[5] so it's no surprise that intoxicated customers and clients can also present challenges in the workplace.

As with other alcohol-related incidents, the safety of employees, clients and others is the primary concern. An employer's duty of care to protect employees by all reasonable means, requires a robust response to any situation where a client or customer is intoxicated and there are potential risks to staff or others. The level of preventative measures required will depend on the industry, level of contact with the public and the likely impact of intoxication. For example, the hospitality industry has long had to deal with incidents involving alcohol and is governed by laws prohibiting the sale of alcohol to someone who is drunk.[6] Licence holders also have the legal right to deny entry to premises on the grounds of intoxication. You will find more information on the 2003 Licensing Act in Chapter 7.

Mr Hui, a twenty-year member of Aspinall's casino in Mayfair, gambled in a private room during a Chinese New Year event. Betting against a signed blank cheque, he reached £500,000 in losses but requested additional credit. The club extended £100,000, which he also lost, after which he drove himself home. Later, Mr Hui stopped the cheque payment, claiming that he was 'blackout drunk' and incapable of signing it. Aspinall's sued him for £589,724 for breach of contract.

The judge ruled that while Aspinall's held the burden of proof, Mr Hui needed to show evidence that casino staff knew or should have known of his intoxication. Mr Hui argued his irrational behaviour and unusual betting amounts were clear signs of impairment, but the casino's evidence showed he had previously been willing to lose £400,000. Since he also drove himself home, it was unlikely staff perceived him as overly intoxicated. The court concluded there was insufficient evidence that the casino should have stopped him from gambling.

(*Aspinall's Club Ltd v Hui*, 2023)

Other sectors such as retail, housing, health and social care and public transport also report dealing with high levels of intoxicated behaviours. Under the influence of alcohol, customers can find it difficult to process information, read body language and maintain emotional stability. Minor misunderstandings can escalate quickly, and the risk of confrontation is increased. In most cases, an intoxicated individual is more of a danger to themselves than others but when a situation escalates it can be distressing to employees and others present.

Fortunately, for most organisations such incidents will be rare, but when they do occur it's best to be prepared. It may be worth arranging training on managing drunken behaviour for employees who are likely to need to deal with intoxication in the course of their work. The following points should be considered when developing a policy and responding to situations.

1. Conduct a thorough risk assessment, which should include considering the potential risk to the intoxicated customers and other members of the public. Risk is dynamic and can change rapidly depending on circumstances so be prepared to review your assessment regularly. Provide an incident book for recording instances of intoxication that are concerning. This will help build a picture of the types and frequency of incidents and can be used to inform future risk assessments.
2. Train staff to recognise signs of intoxication, such as slurred speech, unsteady gait or disruptive behaviour. Make employees aware that erratic behaviour may be due to causes other than alcohol and not to make unfounded assumptions (see *Is it alcohol or something else?*). Provide training on de-escalation techniques to diffuse potentially confrontational situations and model how to respond to incidents effectively. Staff should remain calm, empathic and non-confrontational. When employing door staff check that they are SIA qualified.
3. Arrange rotas to avoid lone working during high-risk times (e.g. Friday and Saturday evenings) or in high-risk areas, such as city centres or client homes. Provide SOS alarms or personal safety devices for lone workers in high-risk sectors. Operate a 'call-in' system for lone workers. There should be a consistent record of staff whereabouts when lone working and employees should always report any deviation from their expected routes or timescales. Advertise your use of CCTV as a deterrent.
4. Document all incidents involving intoxicated customers that lead to a disturbance or require staff intervention. You can use incident reports as training material to help staff learn from past situations and improve the handling of future interactions. Be consistent in your responses and consider contacting law enforcement if there's an immediate risk. Hold other employers accountable for the actions of their employees who are in breach of your alcohol policy (e.g. customers, contractors and other site visitors). Of course, this will require sensitive handling, but clear guidelines in your policy and quick, consistent action will protect your workforce and your organisation's reputation.

Offer debriefing sessions or support to employees who may have been affected by challenging interactions with intoxicated customers or clients. You could also provide resources for employees to seek counselling or assistance if they experience emotional distress because of such incidents.

6.5 DRINK-DRIVING LAWS AND PENALTIES

Under the Health and Safety at Work Act 1974, employers have a duty of care to protect employees and the public from risks associated with workplace activities, including driving on behalf of the organisation. You must take appropriate measures to prevent alcohol-related driving risks. Here in the UK, drink-driving is subject to strict laws and penalties for driving with a breath alcohol concentration of 35µg/100ml breath in England and Wales and 22ugµ/100ml in Scotland. Penalties for drink-driving can include fines, driving bans and imprisonment, depending on the severity of the offence. Employers should emphasise compliance with drink-driving laws and reinforce the organisation's alcohol policy to their employees.

Besides the risk of injury and property damage caused by a drink-driving incident, there may be reputational damage if an employee is convicted of a drink-driving offence. If an employee is charged with a drink-driving offence whilst driving a company vehicle or during their work, the organisation's response should be consistent with the response to other alcohol-related incidents. Follow your policy guidelines and investigation procedures. Once arrested for a drink-driving offence, drivers may legally continue to drive until convicted, but if convicted of driving or attempting to drive whilst above the prescribed limit, they will face a driving ban of at least twelve months (see next for further information on the drink-driving laws).

6.5.1 Prevention

Conduct a comprehensive risk assessment for employees who drive as part of their job. Include factors like work schedules, expectations for client meetings and the possibility of alcohol consumption in certain settings such as hotels. You can prevent situations where employees may be tempted to drink and drive by managing work hours, client entertainment expectations and journey planning to avoid alcohol-related risks.

Educate drivers on units, limits and alcohol processing times (see Chapter 3). This information could be made available in your alcohol policy, or ideally, as part of a mandated alcohol awareness course. Since alcohol can take hours to leave the body, someone who drank late the previous evening could still be over the legal limit while commuting to work the next morning. Even if they're below the limit, residual alcohol may still impair their abilities and contribute to collisions. Over 6,000 people were killed or injured in 2022 in Great Britain as a result of drink-driving incidents,[7] so it's important to send a firm message about the risks of driving under the influence.

Some businesses provide drivers with self-test breathalysers to check that they are alcohol-free before they set out. This gives drivers accountability for their own alcohol consumption and helps to educate them about how long it takes for alcohol to leave the system. Modern self-breathalysers are very accurate, but it is worth emphasising that the purpose is not to check that the individual is under the limit, but that they are completely alcohol-free before driving.

Chapter 3 details the effects that alcohol can have in the short- and long-term, and chapter 4 advises on tackling alcohol problems as soon as they arise as a workplace issue.

BAC is also used as a measurement of intoxication and 80mg alcohol per 100 ml blood[8] is the legal limit Drivers are:[9]

6x more likely to be involved in a fatal crash, with 50–80 mg alcohol per 100ml blood, compared to 0ml
3x more likely to die on the roads, with 20–50mg alcohol per 100ml of blood, compared to 0ml
46% more likely to be at fault in road collisions, if you have 10mg alcohol per 100ml, compared to 0ml

6.6 DRINK-DRIVING LEGISLATION

Under the Road Safety Act, 1967, it is an offence to drive in England and Wales with a breath alcohol level above 35 ug/100 ml breath (80 mg/100 ml blood). In 2014, this limit was lowered in Scotland to a maximum of 22 ug/100 ml breath (50 mg/100 ml blood).

Any uniformed police officer may conduct a breath test on a motorist in the following circumstances:

- After a road traffic collision
- If they suspect that the driver is under the influence of alcohol
- If the driver commits a moving traffic violation.

There are three main drink-driving offences:

1. Driving or attempting to drive a motor vehicle with a breath alcohol concentration above the legal limit or whilst unfit through alcohol consumption
2. Being in charge of a vehicle with a breath alcohol concentration above the legal limit or whilst unfit through alcohol consumption
3. Failing to provide a specimen for analysis when required to do so

Offences numbers 1 and 3 carry a mandatory minimum of a twelve-month driving ban and can also incur a fine, a community order or even up to six months in prison.

Offence 2 *may* result in a ban, although 10 penalty points on the driving licence may be imposed instead. Generally, offences under the Road Safety Act will be tried in Magistrates Court. However, where there has been a collision involving drink-driving, an individual has been injured or there are other aggravating factors, if a custodial sentence exceeding 6 months is likely, the case will be referred to Crown Court.

In some circumstances, a driver will be classed as a high-risk offender and will incur additional sanctions, including a requirement to undertake a DVLA medical before being allowed to apply for reinstatement of their driving licence. These circumstances are:

- Where the breath/blood alcohol concentration exceeds 2.5x the prescribed limit
- Where the driver has committed a previous drink/drug-driving offence within ten years
- Where the driver fails to provide a specimen (of breath, blood or urine) for analysis, on request

Magistrates also have discretion to insist that disqualified drivers re-take a driving test before reinstatement of their licence, although this penalty is not commonplace.

Once convicted of a drink/drug-driving offence, a driver must declare this conviction *when requested* by prospective employers, on insurance applications and travel visas for five years. If convicted of *causing death by careless driving whilst under the influence of alcohol*, the maximum penalty is life imprisonment.

Occupational licence holders, including LGV and PCV drivers convicted of drink- or drug-driving will be referred to the local Traffic Commissioner before the return of their licence. This may result in:

- A written warning (for a first offence or a ban of not more than twelve months)
- An appearance before the Traffic Commissioner and withholding of the licence for a further period of up to three months (first offence, alongside a driving ban of twelve to thirty-six months)
- An appearance before the Traffic Commissioner and withholding of the licence for a period of three to six months (second offence or disqualification of more than thirty-six months)
- Refusal of application for renewal of licence (third offence)

Loss of a driving licence would not also revoke a forklift truck licence and, as long as the forklift operator does not operate the vehicle on a public road (e.g. when moving from one warehouse to another), they can legally continue to work in that role. As a precaution, a risk assessment should be made at this point to determine the likelihood of future impairment whilst at work.

For further information, contact the DVLA at www.gov.uk.

6.6.1 A Drink-Driving Conviction

If an employee is found to have driven with an illegal blood alcohol level, your organisation's response may depend on the employment contract—some require employees to immediately inform their employer of any drink-driving conviction, while others do not address this. A drink-driving conviction will appear on a Disclosure and Barring Service (DBS) check for up to eleven years.

Before conviction, the employee may still drive; however, if they are covered under company insurance, you'll need to inform the insurance provider to see if coverage will continue. While an employer can potentially dismiss an employee with a drink-driving conviction, the fairness of this action depends on various factors, including the job's nature. For instance, dismissal may be unnecessary if the employee's role doesn't involve driving and they can still commute without a vehicle. If driving is only a minor part of the job, alternative arrangements could be explored, such as adjusting duties temporarily.

If driving is essential, consider whether the employee can perform their duties using alternative transportation or by moving them to a different role during the disqualification period. Some employees in driving roles have reached arrangements where family members drive them to appointments, with any extra costs covered by the employee and regular checks by the employer to ensure compliance. In other cases, employees have been reassigned or temporarily given desk duties.

If reassignment or other accommodations aren't feasible, you'll need to initiate the formal disciplinary process. This will include conducting a thorough investigation to confirm the details and, if no reasonable alternatives are available, holding a disciplinary hearing to review the evidence. If dismissal becomes necessary, it would typically be due to a loss of an essential qualification for the role.

The Royal Society for the Prevention of Accidents (ROSPA) advocates regularly assessing driver attitudes towards alcohol, especially new drivers, those who drive the most miles and those using a new type of vehicle. Their guidance adds that employers should require drivers to report collisions and motoring offences to their line manager. This would lead to a discussion about whether the driver needs additional support to prevent repeat occurrences.

NOTES

1 Section 2(1) Mental Capacity Act, 2005.
2 International Labour Organisation. Management of Alcohol- and Drug-Related Issues in the Workplace (https://www.ilo.org/sites/default/files/wcmsp5/groups/public/%40ed_protect/%40protrav/%40safework/documents/normativeinstrument/wcms_107799.pdf).
3 Health and Safety at Work Act, 1974.
4 NHS Digital, 2022. Health Survey for England, 2022, Part 1 (https://digital.nhs.uk/data-and-information/publications/statistical/health-survey-for-england/2022-part-1/adult-drinking).
5 Alcohol Change UK, 2024. Alcohol Statistics (https://alcoholchange.org.uk/alcohol-facts/fact-sheets/alcohol-statistics).
6 Licensing Act 2003.

7 Department for Transport, 2024. Reported Road Casualties in Great Britain Involving Illegal Alcohol Levels: 2022 (https://www.gov.uk/government/statistics/reported-road-casualties-in-great-britain-involving-illegal-alcohol-levels-2022#:~:text=Estimates%20of%20casualties%20in%20collisions,central%20estimate%20of%20300%20fatalities).

8 This limit applies in England, Wales and Northern Ireland. In Scotland the limit is 50mg/100ml blood, or 22μ/100ml breath.

9 Brake, Drink Driving the Risks, the Law, the Limits and the Penalties (https://www.brake.org.uk/get-involved/take-action/mybrake/knowledge-centre/drink-driving).

7 Legal and Ethical Considerations
Alcohol Testing

7.1 INTRODUCTION

By now, you will appreciate that dealing with alcohol issues in the workplace requires a balance to be struck between protecting the organisation, complying with the law and supporting employees. In this chapter, we will explore some of the legal and ethical considerations associated with addressing alcohol-related challenges. We will examine the case for and against workplace alcohol testing, define 'for-cause' and random testing and outline the circumstances and industries where these methods may be appropriate.

QUESTIONS FOR REFLECTION

What is your aim when you initiate workplace alcohol testing?

You may want to consider risk management, creating a deterrent, reducing potential costs, cutting serious incidents, improving wellbeing, ruling out alcohol as a cause of incidents or near misses and so on.

How will you measure the success of your testing regime?

This could be in identifying non-negative tests, indicating that you are identifying those who are in breach, or the opposite: achieving a zero non-negative test rate, suggesting that the testing regime has successfully deterred employees from breaching your policy.

7.2 THE ORIGINS OF ALCOHOL TESTING

7.2.1 1960s

The history of workplace alcohol testing in the UK reflects changing attitudes towards safety and health over recent decades. In October 1967, the Transport Minister, Barbara Castle, introduced the Road Safety Act, making it a criminal offence to drive a motor vehicle with a BAC greater than 80mg/100ml (equivalent to 35µg/100ml breath). Under this new law, police officers were given portable breathalysers and powers to require motorists to provide a breath sample for analysis. Should they fail this test, drivers could be arrested and required to take a further test at a police station, to be administered by a doctor. This could take the form of a urine or blood test.

DOI: 10.1201/9781032648927-8

Refusal to submit to alcohol testing also became a criminal offence. At that time, the maximum penalty for failing a blood test was a fine of up to £100 and up to four months imprisonment.

The law had an immediate effect with drink-drive deaths falling from over 3,000 annually to under 2,000 by the end of 1968. Fifty-five years later, that figure is around 250 deaths per annum on UK roads. The breath and blood limits have remained the same in England and Wales, but breathalysers themselves are unrecognisable from the bulky machines of the 1960s.

Throughout the 1960s and 1970s, concerns about alcohol-related workplace accidents began to surface, especially in industries with higher safety risks, such as mining and public transport. The rise in workplace incidents, such as the Eltham Well Hall rail disaster (see Chapter 2), prompted employers to consider the impact of alcohol misuse on safety and productivity.

7.2.2 1970s

The next piece of legislation on the road to workplace testing was the Misuse of Drugs Act of 1971, allowing for the prosecution of employers who knowingly allowed employees to use illicit substances in the workplace. It was now incumbent on the employer to take measures to reduce health and safety risks from employee drug use and the first workplace drug testing kits were produced in the 1980s. Over the next decade, testing equipment became more precise and more portable—and more in demand. A perceived surge in drug use among young people in the 1980s and 1990s led to increased demand for testing kits, which in turn stimulated research into more sophisticated devices.

The 1974 Health and Safety at Work Act represented another important change in legislation, requiring employers to ensure a safe work environment, though it did not explicitly mention alcohol misuse. By making employers responsible for safety, the Act laid the groundwork for future substance misuse policies.

Employers came to recognise that addressing alcohol-related risks was essential to meeting their safety responsibilities under this law. Increasingly, the focus on workplace health and safety initiatives led to many industries implementing general behavioural and performance standards, particularly in high-risk sectors, although formal policies were still rare. This new awareness of alcohol's impact on safety across other safety-critical industries was bolstered by several fatal accidents linked to alcohol in the 1980s.

7.2.3 1990s

By the 1990s, this heightened awareness of alcohol-related risks led industries such as oil and gas, construction and rail to formalise substance misuse policies. For example, the UK rail industry introduced random alcohol testing following serious incidents, as mandated by the Transport and Works Act 1992, which required workers in safety-critical roles to remain sober while on duty. Due to high safety stakes, the oil and energy sectors also began to introduce alcohol testing policies, incorporating random, post-incident and for-cause testing to minimise hazards from impaired machinery operation and hazardous material handling.

With growing recognition of the impact of alcohol misuse on absenteeism and presenteeism, many employers integrated formal testing into HR policies. Even sectors with lower physical risks began addressing alcohol misuse through EAPs and other support options.

7.2.4 2000s

The 2000s saw policies that mandated testing based on reasonable suspicion or post-incident analysis. The Health and Safety Executive (HSE)[1] and the Trades Union Congress (TUC)[2] both offer guidance to employers, recommending that alcohol testing policies should be transparent, proportionate and legally justified. Employers began including alcohol misuse clauses in contracts to inform employees of potential testing in specific contexts.

During the last fifteen years, random alcohol testing has grown in popularity among high-risk sectors. Industries such as public safety moved towards zero-tolerance policies, where even minimal alcohol levels could lead to disciplinary action. This change aimed to mitigate risks and prevent litigation tied to lapses in safety protocols. Portable breathalysers and mobile testing units became increasingly accurate and affordable, allowing on-site testing with quick results. This made testing accessible even in industries with dispersed workforces, such as transportation. Technology has advanced to the point where devices are available that will prevent the ignition of a vehicle by a driver with a breath alcohol level above the legal limit.

7.2.5 Where Are We Now?

Alongside developments in testing and policies, many employers have developed supportive regimes with help for alcohol issues rather than solely relying on disciplinary measures. Sectors like healthcare and hospitality emphasised early intervention to maintain a workforce that is traditionally vulnerable to alcohol problems. Testing is often perceived as a purely punitive response so best practice today advocates for testing that is proportionate to the role's safety risks rather than a universally implemented measure.

Since the COVID-19 epidemic, with remote work now commonplace, alcohol testing has become less usual for home-based employees. Instead, employers have placed a greater focus on promoting alcohol awareness through virtual wellbeing resources. Alcohol testing is increasingly seen within a larger framework that includes mental health and substance misuse prevention. Many workplaces now offer training to educate employees about managing stress and avoiding substance-related issues.

You will note from the aforementioned history, how closely linked the evolution of workplace testing is to changes in health and safety legislation, and the primary aim of testing regimes is to prevent alcohol-related incidents, with a supplementary aim of acting as a deterrent. The onus is on employers to provide a safe working environment for employees, and this includes limiting the potential risk from workers who are under the influence of any psychoactive substance, not just alcohol.

Of course, this does not absolve the individual of responsibility for their own actions whilst under the influence, and there are several offences linked to drinking and performing activities that may put others at risk. Nevertheless, a testing regime should never be undertaken lightly. By its nature, it is invasive and can impact employer-employee relations and its role in reducing alcohol-related risk should not be overestimated. Good testing policies and practices seek to protect employees and are implemented only when necessary for health and safety and appropriate to the level of risk.

Employers must consider the specific circumstances of their industry and workforce to determine the appropriateness of testing methods. Testing is not inevitable, nor is it necessarily the most effective response to the risks of alcohol harm at work, but it is a recognised means of reaching health and safety compliance. By following established legal precedents and involving all relevant parties in the process, you can strike a balance that safeguards both safety and individual dignity.

7.3 WORKPLACE ALCOHOL TESTING: BALANCING RIGHTS AND SAFETY

Comment

It is good to see evidence that employers are taking health and safety issues more seriously and that there is a change in attitude to personal responsibility for safety. Drug and alcohol misuse poses a challenge to businesses and their ability to effectively manage talent and create sustainable value chains. Implementing a testing regime within an overall policy puts people at the heart of the conversation—quite simply drugs, alcohol and safety-critical tasks such as driving do not mix.

(Dr Karen McDonnell, Occupational Health and Policy Adviser, Royal Society for the Prevention of Accidents [RoSPA])

7.4 COMPLIANCE WITH RELEVANT LAWS AND REGULATIONS

7.4.1 The Equality Act, 2010

An **AUD** as a standalone condition, is not classified as a disability under the Equality Act 2010 (nor under its predecessor, the Disability Discrimination Act, 1995). This means that employees who misuse alcohol do not automatically receive protections afforded to other disabilities or impairments. However, if alcohol misuse results in another mental or physical impairment, such as depression, liver disease or anxiety, that condition *can* qualify as a disability if it substantially affects the employee's ability to perform everyday activities over a long-term period.

When an employee has a condition classified as a disability (such as depression stemming from alcohol misuse), the Equality Act protects them from both direct and indirect discrimination. For example, if an employer terminates or penalises an employee solely because of a disability-related performance issue without considering reasonable accommodations, the employer could be acting unlawfully under the Act. Employees are also protected from harassment or victimisation due to a disability or alcohol-related condition.

> An employee argued that her dismissal was unfair and discriminatory due to a disability, as her alcohol addiction had resulted in depression. Under the Disability Discrimination Act (DDA), her condition would not typically qualify as a disability. However, under broader regulations, her depression could be considered a disability regardless of its cause. On appeal, it was determined that the tribunal should not have focused on alcohol dependence as the cause of her impairment but instead should have considered her depression itself to assess whether discrimination occurred.
>
> (*Power v Panasonic UK Ltd*, 2003)

The Equality Act requires employers to make reasonable adjustments for employees with recognised disabilities. For those with alcohol-related secondary conditions, adjustments could include:

- Flexible working hours to attend medical or other therapeutic appointments.
- Modified work responsibilities to account for any temporary performance impairments related to treatment or recovery.
- Temporary remote work or shift changes to accommodate the employee's treatment needs.
- Provision of equipment or modifications that might ease stress factors at work (e.g. ergonomic changes).

Employers can arrange referrals to OH providers if an employee's **AUD** has led to a secondary condition that impacts their work performance. OH specialists can assess the employee's health in a workplace context and can recommend workplace adjustments without breaching employee privacy.

7.4.2 Employee Privacy and Rights

Any information about an employee's alcohol-related health condition must be treated as confidential and must follow **GDPR** guidelines regarding sensitive health information. Any personal information related to the condition or to alcohol test results should be securely stored and only disclosed to necessary personnel. Employees must also be informed of any policies that may affect them, including the right to refuse certain medical assessments (such as alcohol testing), though refusal could have workplace consequences in safety-critical roles.

7.4.3 Considerations for Disciplinary Action or Dismissal

UK employment law emphasises that any disciplinary process for alcohol-related misconduct or capability concerns must be fair and transparent. Employers need to demonstrate that they have followed due process:

1. Before acting, employers should collect evidence, which might include witness accounts or behavioural observations, to substantiate any claims of impaired performance or conduct.
2. In cases where an employee is struggling due to alcohol misuse, employers should consider offering support or alternatives before proceeding to dismissal, especially if the employee is in recovery or has a recognised disability.
3. If the employee's role requires driving or safety-sensitive responsibilities, dismissal may be justifiable if the employee cannot safely perform their duties due to alcohol misuse. However, employers must first consider potential alternative arrangements, such as temporary reassignment, allowing unpaid leave or alternative transport arrangements if feasible (e.g. a designated driver in case of a driving ban).
4. If misconduct has been linked to alcohol misuse, employers should conduct a fair investigation so that decisions are based on factual, evidence-based grounds. This could include looking at the employee's past performance record, the nature of the incident and any reasonable steps the employee has taken to address their alcohol misuse. If a decision to dismiss is made, a final failsafe check, the Burchell Test, can be applied.

THE BURCHELL TEST

This test is a legal principle used in Employment Tribunals that evaluates whether an employer acted reasonably in dismissing an employee for alleged misconduct by examining three key questions:

Did the employer sincerely believe the employee was guilty of misconduct?
Did the employer have reasonable grounds for this belief?
Was this belief formed after a fair investigation?

It's important to understand that the test does not require the employee to be proven guilty of misconduct. Even if the employee later demonstrates their innocence, the employer's decision may still be deemed reasonable if they met the aforementioned criteria. If all three questions are answered 'yes', there is a strong likelihood that a Tribunal will find the employer's decision to dismiss was fair.

If an employer suspects an employee of misconduct, they must follow proper procedures, including collecting evidence, reviewing relevant documents, speaking with the employee and providing written statements, interviewing any witnesses and ensuring a thorough and reasonable investigation into the alleged misconduct.

7.5 HEALTH AND SAFETY OBLIGATIONS

7.5.1 The Health and Safety at Work Act, 1974

This act is a cornerstone of UK health and safety law, requiring employers to ensure, so far as is reasonably practicable, the health, safety and welfare of all employees. This duty includes identifying and mitigating any risks that may arise due to alcohol use, that could impair judgement, slow reaction times and increase the likelihood of accidents, especially in high-risk roles.

Beyond employee safety, the act also requires employers to protect anyone affected by an organisation's operations, such as contractors, clients and the general public. This duty is particularly relevant in industries where alcohol impairment could endanger others, and negligence could *potentially* leave the employer open to prosecution under the 2007 Corporate Manslaughter and Corporate Homicide Act.

In high-risk roles, an employer's responsibility to promote safety may lead to additional protocols including 'reasonable suspicion' testing if there are signs of impairment. However, such testing must align with privacy regulations, and employers are encouraged to consult employees and union representatives when introducing such measures.

Employers must conduct risk assessments that identify and evaluate potential hazards. This includes evaluating specific job roles where impairment could create dangerous situations, such as operating heavy machinery, driving vehicles or managing executive tasks. It's important to look at both immediate risks and the potential impact on workplace productivity.

You should document risk assessments, training, policy reviews and any incidents related to alcohol. Detailed records serve as proof of compliance and demonstrate a commitment to workplace safety if an incident occurs.

The Advisory, Conciliation and Arbitration Service (ACAS) and the Health and Safety Executive (HSE) provide guidance on best practices in handling alcohol misuse. They emphasise that:

Testing or disciplinary measures should be proportionate and in line with company policy.

Policies should be transparent and agreed upon, ideally in consultation with staff or unions, ensuring all parties understand the implications.

Employers should focus on supportive, non-punitive approaches where possible, such as offering EAP resources or access to counselling services.

7.5.2 The General Data Protection Regulation

Under the **General Data Protection Regulation (GDPR)** and the Data Protection Act 2018, employers must demonstrate a clear, lawful basis for processing alcohol-related information data, which often involves ensuring the data is directly relevant to workplace health and safety or required by legal compliance. Employers also need to satisfy a lawful basis for gathering data (such as legitimate interest or legal obligation). This helps protect employees' rights by ensuring sensitive data is processed only when necessary and within legal parameters. Confidentiality may be breached under specific legal circumstances, such as when there is a risk of harm to the employee or others or when required by a court order. Even in such cases, the breach should be limited to what is *necessary*.

When handling health-related data (including alcohol test results), **GDPR** requires that employees give explicit, informed consent for data collection except where the data processing is strictly necessary for legal or safety compliance. This consent should be freely given, well-documented and revocable. For example, in cases where alcohol testing is required to guard workplace safety, employees should be fully informed about what the test entails, how the results will be used and who will have access to the data. They must also be informed about their data rights, including how long the data will be stored, the purpose of data collection and how it will be processed. If consent is required, employees must have a clear understanding that they have the right to refuse.

Employers should collect only the information necessary for legitimate purposes and retain it for as short a period as possible. For example, if an employee undergoes alcohol testing following a workplace incident, the data should be retained only as long as necessary for investigative or compliance purposes, then securely deleted in line with the employer's data retention policy. Employees can request the deletion of their health information if the data is no longer necessary or if the processing was based on consent that has since been withdrawn. Employers must have policies in place to respond to these requests and securely erase any data that the criteria outlined in **GDPR.**

The Information Commissioner's Office (ICO) and ACAS provide guidance for handling health-related data, recommending that employers adopt clear policies and secure employee consent where required.

7.5.3 Workplace Alcohol Testing

[Is the test] *relevant with regard to its aim, does the test do what it is supposed to do? If improving workplace safety is the goal, the basic question about relevance is whether testing on alcohol leads to a safer workplace. A basic performance test may be a better solution.*

[is the test] *proportional to the underlying interests. Is testing a sample a case of shooting at a mosquito with a cannon ball or is the test in balance with the interests of colleagues, the company and the general public? Taking a urine sample from a pilot on duty when under suspicion of being under influence vis-à-vis the safety interest of the passengers and society, seems an easy case, but a urine sample taken from a randomly chosen secretary or administrator is much more complex.*

[is the test] *the least extreme and intrusive method available. Sampling blood is more intrusive than urine, which in turn is more intrusive than breath.*

Doing a virtual performance test is less intrusive than taking a sample of any kind from a person. The least intrusive test should be used.

(Tobutt, C. Alcohol at Work. Managing Alcohol Problems and Issues in the Workplace)

7.5.4 When Workplace Alcohol Testing Is Appropriate

Workplace alcohol testing is generally considered appropriate in high-risk, safety-sensitive industries where impairment can directly endanger lives or business operations. Several sectors employ testing protocols for compliance, to act as a deterrent and to ensure that employees responsible for safety remain alert and unimpaired. We established in Chapter 1 that there are workplace risk factors for alcohol problems, including working in physically dangerous conditions and exposure to personal safety risks—often conditions found in these high-risk occupations.

Testing may also be warranted following accidents or incidents at work, where it is crucial to determine if alcohol played a role in the event, helping employers address potential safety issues proactively. Random testing in high-risk roles is also seen as appropriate, particularly for positions such as air traffic controllers and nuclear power operators, where any lapse in focus or judgement could be catastrophic. Testing also provides objective, evidence-based data that can support disciplinary actions in cases of suspected misconduct, helping to establish that any repercussions are fair and based on clear policies.

However, there are ethical concerns with testing, particularly surrounding issues of privacy and dignity. Critics argue that alcohol testing—especially when applied without clear evidence or suspicion—can be intrusive and degrading for employees. Employees may perceive alcohol testing as punitive, leading to resistance, decreased morale and a strained employer-employee relationship. Testing without informed consent is also problematic, as it may be considered a violation of privacy rights and can be legally contentious. Alcohol testing can also create stigma, discouraging employees from seeking help for alcohol issues out of fear of repercussions for their employment.

Moreover, relying solely on testing as a response to potential alcohol misuse—without supporting values and practices that encourage open, non-judgemental dialogue, backed by support and access to treatment options—fails to address underlying issues that contribute to alcohol-related impairment. These limitations highlight the importance of a considered approach to implementing a testing regime.

Alcohol testing regimes are costly and inconvenient at best, and in isolation will not prevent problems. Where there is no legitimate safety or business-critical imperative, testing may not be the solution and other means of reducing risk should be implemented first. Hopefully, your employee consultation, risk assessment and monitoring of the effectiveness of your alcohol policy will help you to identify what gaps need to be filled and whether testing is appropriate.

7.5.5 Instituting a Workplace Alcohol Testing Regime

If you've made the decision to test, you will need to follow strict procedural guidelines on how to proceed. Testing policies should adhere to ACAS recommendations to and give thought to employee rights and legal compliance. Although in a recent survey,[3] over 75% of respondents claim that they would comply with a testing policy and can see the benefits, this doesn't always play out in practice.

Consultation with staff, managers, unions and other stakeholders should begin at an early stage. HR professionals and management will play a major role in developing, communicating and enforcing policies that align with legal standards and ethical practices. The development of these policies requires compliance with relevant laws and regulations while taking into account the specific needs of the industry and workforce. Once a policy is developed, both HR and management teams must undergo thorough training on its details, particularly on when and how testing should be administered.

Training should give managers the skills to work proactively with your testing policy and understand the need to respect employee rights, which is vital to maintaining good relations and minimising the risk of legal issues. Effective communication of the policy to all employees is equally critical; HR should clearly inform staff about the procedures, expectations and their rights and responsibilities under the policy.

The oversight of testing procedures is another responsibility of HR and management. You must make sure that these procedures are conducted with consistency and fairness, aligned with the standards established in the policy. When it comes to conducting tests, this can be completed in-house by trained staff (e.g. for administering a breathalyser test) or externally by professional testing companies. Testing technicians need to demonstrate accuracy and neutrality so employing an external organisation can be beneficial, particularly where more invasive testing methods, such as saliva or urine sampling are required. Testing companies offer a range of solutions from just the supply of breathalysers to policy and protocol development, urine and saliva test administration, laboratory work and legal advice.

Testing is a skilled task. Technicians are required to handle testing equipment, interpret results accurately and meet all legal compliance requirements. Impartiality in random testing is particularly important, as it prevents potential bias and maintains the integrity of the testing process. It's also imperative to maintain a documented chain of custody for all collected bodily samples for to prevent tampering and establish the reliability of the test results. You will need to communicate the test outcomes to the relevant parties as quickly as possible. In particular, informing the employee involved and any authorities if required. This will help to embed the testing regime as a fair and necessary process. This structured approach to testing, combined with a clear reporting process, is essential for upholding both organisational and legal standards. Your technician will be able to advise on the most appropriate testing methods to use.

7.5.6 Sample Collection Methods

TABLE 7.1
Methods of Collecting Samples When Testing for Alcohol Use in the Workplace

Test	Procedure	Role of technician
Breath alcohol testing (breathalyser)	Employees blow into a breathalyser device, which measures the concentration of alcohol in their breath. The results are typically immediate.	A trained and certified technician administers the test and interprets the results.
Saliva testing	A swab is used to collect saliva from the employee's mouth, which is then tested for the presence of alcohol. Results are typically available within minutes.	A trained technician collects the sample and performs the test.
Urine testing	A urine sample is collected from the employee and analysed for alcohol content. Results may take longer to obtain compared to breath or saliva testing.	A trained laboratory technician handles the sample and performs the analysis.

For accuracy, it's vital to use reliable methods and well-maintained equipment that reflect the true alcohol content in an individual's system. Calibrate breathalysers regularly and dispose of out-of-date swabs or solutions. This precision helps avoid false positives, which could lead to unwarranted consequences. By standardising testing procedures across the business, you can mitigate risks of bias or discrimination and reduce the likelihood of false positive test results.

Respecting employees' privacy and rights throughout the testing process is also vital both legally and ethically. You can avoid bias by having trained, certified professionals conduct tests and interpret results which also helps to minimise errors. This will also ensure that each stage of the testing process is properly documented and tracked.

WORKPLACE ALCOHOL TESTING REGIMES

For-cause testing

For-cause testing, also known as reasonable suspicion testing, occurs when there is a specific reason to believe that an employee may be under the influence of alcohol. This testing is typically based on observable behaviours or signs of impairment. Testing may also be undertaken after an incident that causes or has the potential to cause harm.

Random testing

Random alcohol testing involves selecting employees for testing without any specific suspicion or reason. The selection process should be completely random and unbiased, using a random number generator, for example. Usually around 10%–20% of the workforce will be tested in a twelve-month period. Being random, some employees may find themselves regularly subjected to testing, whilst others are never tested.

7.5.7 Permission to Test

If permission to test is written into the employee's contract of employment, then consent is implied and a refusal to test on request may be grounds for disciplinary action. If there is no contractual obligation, testing is deemed to be voluntary, and a refusal should not in itself result in a disciplinary approach. If there is reasonable suspicion that the employee is under the influence of alcohol; however, this should trigger an investigation and if the suspicion is deemed to be founded, there may be disciplinary consequences. In most cases, failure to provide a sample for testing or a non-negative test result when under contractual obligation could be deemed gross misconduct.

Testing following workplace incidents or accidents, particularly where impairment is suspected, should be conducted immediately to ensure reliability. If there are observable behaviour or performance issues suggesting impairment, testing may be conducted based on reasonable suspicion. Employees must provide informed consent prior to testing so you will need to demonstrate that they understand the process, their rights and the potential consequences of refusal. If an employee refuses or tests positive (technically called non-negative), employers must outline possible repercussions, including disciplinary actions and the right to representation.

Employees should be informed of a process to appeal or the right to challenge results they believe may be inaccurate, with a defined timeline for resolution. Consequences of non-negative tests should be communicated, including disciplinary actions or follow-up testing if necessary. As described earlier, where appropriate, reasonable accommodations should be considered for employees with alcohol-related issues that qualify as disabilities.

CASE STUDY

A longstanding employee was dismissed after admitting to consuming 'a pint of beer' before starting a 2 pm shift. A breath test was then requested, and the employee was warned that if he was found to be over the limit set in the company's alcohol policy (equivalent to the drink-driving limit) or if he refused to provide a sample, he would face disciplinary proceedings. The employee repeatedly failed to complete the test. The company doctor examined the employee and consulted his medical records and could find no medical reason for failure to complete the test. The dismissal was upheld at the tribunal as the organisation had followed proper procedure in seeking a test and a refusal without reasonable cause was deemed gross misconduct.

(*Vaghadaia v Metalastic Ltd*, 2000)

7.5.8 Screening and Evidential Testing

Breathalyser tests can be divided into two categories: screening and evidential tests. Screening tests are designed for quick sampling and their main benefit is their efficiency, with the capability to screen several people within a short time frame. These tests provide simple results, either a 'pass' (negative result) or 'fail'. However, a 'fail', normally referred to as 'non-negative', does not definitively indicate alcohol use, as various environmental substances could trigger a positive result.

For cases where the screening test result is non-negative, a follow-up with an evidential test is necessary. Evidential tests require the use of a mouthpiece, which helps to eliminate external contaminants' influence on the results.

7.5.9 A Non-Negative Test Result

Non-negative tests are relatively rare, with less than 4% of drug and alcohol tests registering a non-negative result.[4] Point-of-care breathalysers are more than 99% accurate but in a very small minority of cases, false positives can occur.

During the period between a non-negative result and confirmation or evidential testing, your employee will benefit from your assurances that the result will be kept confidential and should be encouraged not to disclose information about the test to colleagues. If appropriate, they can be assigned to other duties whilst an investigation is underway. If an employee discloses an alcohol problem, admits to alcohol use or a secondary testing method confirms a positive test result, your disciplinary and support measures should be triggered, depending on whether you wish to proceed via the conduct or capability route.

7.5.10 Disclosure of Alcohol Problems

Employees should always be encouraged to disclose alcohol problems to a trusted manager, HR professional or health champion. But this is hard. In an alco-centric society drinking is the norm, and any problems can be dismissed as one-offs, stress-related, circumstantial or 'social'. Against this backdrop, recognising an issue, let alone acknowledging it and seeking help is notoriously difficult. Current estimates suggest that only around 15% of adults with an AUD are engaged with recovery services[5] so when we advise employees to seek help from their employer in order to access support and to avoid disciplinary action if they are later found to be impaired by alcohol, we're asking a lot. Your actions in developing a fair policy and implementing supportive measures will help, but there may be occasions when, despite your best efforts, an employee with an **AUD** returns a non-negative test.

A non-negative test result is not necessarily indicative of an **AUD** and may suggest reckless or careless action on the part of the employee, but in some cases, a colleague's drinking may already be an open secret and a non-negative test is reflective of their workplace performance or attendance. Occasionally, an employee who has shown no prior indicative behaviours will return a non-negative test and admit to an alcohol problem at this stage.

The circumstances around the test result—random, for-cause, post-incident—and your employees' disciplinary record, length of service and other factors will need to be considered once a positive result has been confirmed. If the employee wishes to challenge the result, an appeal process (and possibly conflict resolution protocols) will need to be instigated, but if not, very careful consideration will need to be given to the consequences for the employee. Your policies (alcohol and disciplinary) should outline the range of sanctions that can be applied to a policy breach, up to and including dismissal. Taking a compassionate, supportive stance at what will undoubtedly be a low point for your employee can profoundly impact their recovery journey.

7.5.11 Fair Recruitment Practices

Pre-employment inquiries should steer clear of invasive questions related to an applicant's personal habits, including alcohol use, as these can be seen as discriminatory and discourage otherwise qualified individuals from applying. Instead, recruitment efforts should emphasise the essential skills, qualifications and competencies relevant to the job. Assumptions about an applicant's abilities should not be based on any personal challenges they might face, including those related to alcohol. Employers should also prepare to provide reasonable accommodations, such as extending time for assessments or adapting interview formats, if a candidate has an alcohol-related disability.

The key legislation governing alcohol and work is summarised next.

7.6 CRIMINAL LEGISLATION

TABLE 7.2
Table of Legislation and Offences Relating to Intoxication When in Charge of Motorised Vehicles

Act	Offences	
Road Safety Act, 1967, s.1, s.2	Driving or being in charge of a motorised vehicle with a BAC above the prescribed limit; offence to refuse a breath test without reasonable grounds	
Railways and Transport Safety Act, 2003 s.78/s.92 Up to 2 years imprisonment or fine	Shipping Offence when on duty or on call to have ability to carry out his duty impaired by the use of drink or drugs	Aviation Offence to perform an aviation function (or ancillary function) at a time when ability to perform the function is impaired because of drink or drugs
Road Traffic Act 1988 s.4/s.5 Drink-Drive First Offence Maximum penalty: 6 months prison sentence, driving ban of 12–36 months, unlimited fine	It is an offence for a person to be in charge of a motor vehicle on a road or other public place with excess alcohol in his breath or any amount of certain controlled substances in blood or urine as evidenced by a certificate of analysis or statement.	It is an offence for a person to be in charge of a mechanically propelled vehicle on a road or other public place while unfit through drink or drugs.
Transport and works Act, 1992 s.27 Maximum 6 months prison sentence, fine up to £5,000	It is an offence to work as a driver, guard, conductor or signalman or in any other capacity in which one can control or affect the movement of a vehicle if unfit to carry out that work through drugs or alcohol	It is an offence to work in a maintenance capacity or as a supervisor of or look-out for, persons working in a maintenance capacity if unfit to carry out that work through drugs or alcohol
	If a person commits an offence under Section 27 in the course of his employment with a person other than the responsible operator, his employer shall also be guilty of an offence. No offence is committed if the employer has exercised all due diligence to prevent the commission on the transport system by any of his employees of any offence under Section 27.	

7.7 EMPLOYMENT LEGISLATION

TABLE 7.3
Table of Employment Legislation Relating to Alcohol and Other Drug Use

Act	Relevance
Data Protection Act 2018	The sharing of employee data, including drug or alcohol use, past problems with substances etc., needs to be in line with this legislation and should be restricted to those who need to know.
Employment Rights Act 1996	Substance use, where it impacts the performance of the employee or the company or puts employee, colleagues or wider society at risk may be cause for dismissal.
Equality Act, 2010 (subsequent to the Disability Discrimination Act, 1995)	Dependence on drugs and alcohol is not covered by this act. A physical or mental impairment caused as a result of addiction is covered, e.g. liver disease. Further, the addiction may be a symptom/side effect of another medical condition, e.g. depression. However, substance misuse problems should be dealt with as health conditions.
Human Rights Act 1998	Relevant when considering testing for drugs/alcohol. Employee has right to 'respect for private and family life'. This may be outweighed where the employee carries out a safety or business critical role.

7.8 HEALTH AND SAFETY LEGISLATION

TABLE 7.4
Table of Health and Safety in the Workplace Legislation Relevant to Use of Alcohol And Other Drugs in the Workplace

Act	Relevance
Health and Safety at Work Act, 1974	Employers have a duty to ensure, as far as is reasonably practicable, the health, safety and welfare at work of their employees. Employees have a duty to take reasonable care for their own health and safety and that of others who may be affected by their acts or omissions at work. Employers who do not deal with employees who are under the influence of substances at work may risk prosecution.
Management of Health and Safety at Work Regs, 1999	Employers need to make a 'suitable and sufficient' assessment of risks to the health and safety of employees to which they are exposed whilst at work. Employers are also required to act on any identified risk, e.g. if aware that an employee is intoxicated.
Corporate Manslaughter and Corporate Homicide Act 2007	Whilst alcohol is not explicitly mentioned, employers can be found guilty if failure to implement adequate policies and procedures leads to fatal consequences.

NOTES

1 Health & Safety Executive, 1996. Don't Mix it. A Guide for Employers on Alcohol at Work (https://www.ias.org.uk/uploads/pdf/In%20the%20Workplace/indg240.pdf).

2 Trades Unions Council, 2019. Drugs and Alcohol in the Workplace. Guidance for Workplace Representatives (https://www.tuc.org.uk/sites/default/files/drugsalcoholinworkplace.pdf).
3 Drager UK Safety at Work Report, 2023 (https://www.draeger.com/en_uk/Safety/The-Draeger-Safety-and-Health-at-Work-Report).
4 UK Drug Testing, 2022. Random Drug Testing at Work (Blog) (https://www.ukdrugtesting.co.uk/blogs/news/random-drug-testing-at-work?srsltid=AfmBOopDglBLa_1tgWbv5Z2jBEYEJvcUINXEHO0qSwjGsZ9RKX3DMG7v).
5 Public Health England, 2021. Adult Substance Misuse Treatment Statistics, 2020 to 2021: Report (https://www.gov.uk/government/statistics/substance-misuse-treatment-for-adults-statistics-2020-to-2021/adult-substance-misuse-treatment-statistics-2020-to-2021-report#treatment-interventions).

8 Support and Treatment for Alcohol Use Disorders

8.1 INTRODUCTION

In the UK, companies have a growing responsibility to not only identify and manage alcohol-related risks but also create an environment where employees feel supported in seeking help without stigma or fear of discrimination. Providing support and treatment for employees with an **AUD** is a necessary component of creating a safe and happy workplace. Organisations increasingly recognise that by addressing alcohol-related challenges in a structured and compassionate way, they can improve both the outcomes for individual employees and the workplace environment.

In previous chapters, we determined that effective support begins with awareness of the complexities of addiction and an understanding of how alcohol can impact the workplace. By training OH and HR managers to notice the signs of alcohol misuse and developing clear protocols for how to approach the employee, you have set the scene for responding to disclosures of alcohol problems in a responsible and supportive manner. This chapter examines how you can navigate these sensitive situations and develop strategies for building a supportive framework—from early help to offering accessible treatment pathways and appropriate accommodations.

8.2 WHEN DRINKING BECOMES A PROBLEM

In Chapter 4, we looked at the most common, effective and well-evaluated response to high-risk drinking—screening and brief advice. For the vast majority of the more than 10 million people in the UK who regularly drink at above the recommended maximum, this is generally sufficient to highlight the risks and set the drinker on a different path. If caught early, drinking patterns can be easily adapted for a safer lifestyle, which is why your role as an employer is so important. By establishing open, non-judgemental discussions, you can make employees more likely to seek assistance, reducing the chance of alcohol-related issues in future.

For some individuals, though, what may start as drinking in excess to deal with stresses or trauma or life challenges can quickly escalate to feeling out of control. Many people can continue for years, decades even, without acknowledging to themselves or others that their drinking has become problematic. An alcohol problem isn't easy to define—it's not about how many units are being consumed or what type of alcohol or how frequently, so there are no quantitative markers to delineate that *this a problem, but that isn't*. A 2017 survey[1] found that nearly 60% of respondents who

DOI: 10.1201/9781032648927-9

drink in excess of the recommended guidelines had no intention to make changes to their drinking and just 6% felt motivated to cut down.

Your wellbeing strategies and policies around alcohol will contribute to providing a landscape that makes recognising and responding to alcohol-related problem, a less rocky and erratic process.

AUDs are messy things. Recovery is rarely linear, and relapse rates are high. What's more, with less than 20% of individuals who have an AUD accessing treatment, and around half dropping out within a few weeks of starting a treatment regime, supporting someone through this process can be resource-intensive and frustrating. Still, losing a valued employee, filling a void with temporary staff, re-hiring and training a new employee can be equally so. Putting procedures in place that seek to support employees who are struggling with their drinking is a compassionate and ethically sound practice, but it will also need to be boundaried and standardised across the organisation.

8.2.1 A Conduct Issue

*A **conduct issue** arises when an employee's actions breach company policies, practices or ethos due to alcohol misuse. This includes behaviours such as frequent unexplained absences or patterns of sick leave due to alcohol-related issues; disregarding company rules or refusing to follow instructions or engaging in disputes or conflicts with coworkers as a result of alcohol-related behaviour.*

(Chapter 4)

An AUD often comes to light when an employee's conduct has contributed to an incident or near miss or their behaviours require a disciplinary response. Sanctions do not dissuade an individual from drinking once they have developed an addiction. Whilst it is necessary to enforce your disciplinary procedures commensurate with the infraction, this is unlikely to have a positive effect on your employee's alcohol use. Quite the reverse in fact, as negative emotional states are often a trigger for further bouts of drinking.

Once you've established that a conduct issue is alcohol-related, you'll need to consider what action you will take to prevent a recurrence. Your response may be advisory, leaving your employee to choose whether they wish to access additional support or not or mandatory, in which case their adherence to any requirements will be monitored.

8.2.2 Advisory Responses

Where the infringement is minor or it's accepted that an incident is out of character and may not be due to an underlying alcohol problem, an advisory response may be the best response. This gives the employee an opportunity to reflect on their circumstances and consider a range of options. A supportive manager or wellbeing champion can offer to help the employee to evaluate whether they would benefit from additional support. This would be a good opportunity to deliver the AUDIT tool (see Chapter 4) and offer brief advice. Other suggested courses of action for the employee to consider include contacting their GP, seeking support from the company's **EAP**, alcohol coaching, self-help or talking therapies.

Any advice about help-seeking should be recorded and given to the employee in writing. You should also hold regular follow-ups to establish whether the concerning behaviours have been resolved. If the issues continue, you may need to consider a mandated response.

8.2.3 A Disclosure

In some instances, an employee may make an unprompted disclosure about their difficulties with alcohol. This is often in response to changes in policies or after successful wellbeing initiatives that focus on alcohol. It's important to acknowledge what a brave action this is. There is still a huge stigma around alcohol problems and open discussion of difficulties remains largely taboo. If an employee has the confidence to ask for help for an AUD, it is a credit to your workplace.

Disclosures also require an advisory response. If the employee would find it helpful, then monitoring of their engagement and regular check-ins can be arranged. Unlike situations in which there is a business imperative to resolve an AUD, a voluntary disclosure should not trigger a disciplinary response. Rather, treat it just as you would any other health issue. Keep all information confidential, explore with the employee what support or workplace adjustments they may need, check any safety considerations and make a risk assessment, then carry on as before. If the employee's difficulties worsen and this begins to affect their attendance or productivity, you may need to switch to a more proactive stance and take steps to resolve a capability problem.

8.2.4 A Capability Issue

> *A **capability issue** arises when an employee's ability to perform their contractual duties is compromised due to alcohol misuse. This may include a noticeable decline in an employee's output and performance; an employee's inability to meet the expectations outlined in their job description or when an employee's alcohol misuse poses safety risks to themselves or others in the workplace.*
>
> (Chapter 4)

Deficits in capability due to alcohol use require a systematic approach so that the employee's performance can managed, and any improvements can be monitored. Set timescales and SMART goals so that both employee and employer can gauge whether improvements are in line with the targets set. Where there is no or inadequate improvement, a graduated response can be implemented that is supportive to the employee but also escalates the intervention level as appropriate. Again, you may wish to begin with an advisory approach, but it would be appropriate in the case of a capability deficit to respond with mandating positive action from the employee to improve performance or attendance.

The nature and level of support required can be negotiated with your employee. Setting careful parameters around the improvement that you wish to see in the workplace or the level of engagement that is required with a treatment programme is essential. All mandated options should be thoroughly discussed with and agreed upon by the employee and all agreements should be in the form of a written contract, signed by both management (or HR/OH) and the employee. In this way, any breaches can be easily identified.

8.2.5 Gross Misconduct

Where an employee is to be disciplined for an alcohol-related breach that constitutes gross misconduct, a requirement to attend alcohol treatment could be an alternative to dismissal. In this case, an HR representative, the employee and their line manager (and other relevant stakeholders) should meet to determine what supportive but compulsory treatment option should be followed by the employee. Coercive treatment can be an effective way of helping an individual to accept that an intervention is necessary. In many cases, being in a recovery environment and removing the element of choice can help the individual to engage. It's also an opportunity to dispel any myths they may believe about what treatment entails. Whilst *'You can lead a horse to water, but you can't make it drink'*, if you can get the horse in front of the trough, it may realise that it's thirsty and drink of its own accord.

> *Movie versions of rock bottoms are dramatic. The character nearly kills someone in a car crash. They get caught in bed with someone else. They wake up and find a tiger reclining in their bathroom.*
>
> *The reality of most drinkers' stories is that the last day drinking is more of a whimper than a bang. It's a slow, insidious undoing.*
>
> *(Catherine Gray, The Unexpected Joy of Being Sober)*

8.3 SUPPORTING EMPLOYEES WITH ALCOHOL PROBLEMS

8.3.1 Initial Assessment

Supporting employees with alcohol problems may seem daunting but, just as with any other OH or safety issue, taking the time to develop a well-thought-out, supportive plan that addresses both the presenting difficulties and underlying causes, will pay dividends. I say again, your managers have the skills to respond effectively, it's often a psychological block about dealing with alcohol issues that can get in the way. So any resources, training or other tools that you can equip them with, will increase their confidence going forward.

Your employee, their line manager and an OH or HR professional all have a role in determining how to proceed with support. Whilst it is up to the organisation to decide on the level and duration of support that you will offer, it is helpful to understand the employee's circumstances, their role in the company and their willingness and ability to engage with the support available. This will enable you to tailor your approach to the employee's needs and be certain that business aims aren't compromised.

By actively involving your employees in your initial assessment, you will be better equipped to collaboratively produce a recovery plan that they have confidence in, and you will be better able to identify possible pitfalls. Your assessment should also build in contingencies for when difficulties arise, for example due to changes in business practices or employee circumstances.

You will also need to consider how information will be shared both internally and from external providers and access to sensitive information should be limited to only those directly involved in the case. Conversations about alcohol-related issues should be conducted privately and respectfully. Where alcohol treatment is required as a response to a policy breach, it is usual for the organisation to receive updates about progress from any external providers. This should be restricted to determining whether the employee is meeting agreed engagement, attendance and compliance goals and should not include information about therapeutic conversations.

QUESTIONS FOR CONSIDERATION INCLUDE:

Will the employee continue to work during their treatment period? If so, will they be able to carry out normal duties? Do you need to consider fitness to drive or operate machinery or make business-critical decisions?

What barriers might there be to the successful completion of treatment (such as availability of appropriate support, lack of social support or disciplinary matters).

Are there any training needs? This may be additional safety briefings or manager training in how to support employees.

8.3.2 Plan of Action

To determine the appropriate level of support, you'll need to evaluate your employee's progress and willingness to engage, as well as their ability to fulfil job responsibilities while maintaining their recovery aims. Employers should collaborate with professionals, involved in the recovery process, such as counsellors or EAP specialists, to assess the employee's situation and recommend tailored interventions. It is vital to establish measurable benchmarks for improvement, such as consistent attendance or participation in treatment or improved performance. These goals must be SMART and should be communicated clearly to the employees, ensuring they understand the expectations and consequences.

8.3.3 Reasonable Accommodations

Employers are legally obligated under the Equality Act 2010 to make reasonable accommodations for employees with alcohol-related problems when they are linked to underlying physical or mental health conditions. These accommodations might include for example, offering flexible work hours to allow the employee to attend therapy or counselling sessions or temporarily modifying job responsibilities. This can be especially beneficial in high-pressure roles, to alleviate stress and create a more manageable workload while the employee focuses on recovery. In some cases, you may wish to offer a temporary leave of absence to enable the individual to participate in intensive rehabilitation or treatment programmes without the added pressure of work obligations.

Additionally, you could assign a mentor or coach within the organisation to provide the employee with a consistent source of guidance and encouragement. This mentor might be a colleague, a wellbeing champion or a manager who can offer ongoing advice and serve as a point of contact during the recovery process.

To implement these accommodations well, it is important to engage in open and interactive dialogue with the employee. This conversation should explore their specific needs and how these align with their treatment plan, making sure that the accommodations are both practical and meaningful. Once agreed upon, accommodations should be clearly documented to provide a mutual understanding of the terms, while also safeguarding the employer's compliance with legal and policy standards.

Accommodations should not be static; they require continuous reassessment to remain relevant as the employee progresses through recovery. Regular check-ins with the employee can help determine whether adjustments are necessary, ensuring that support evolves in line with their changing circumstances. You will need to demonstrate a commitment to the employee's wellbeing *and* maintain good boundaries and accountability to give their recovery the best possible chance of success.

Elements of a good recovery plan are included in the following checklist:

TABLE 8.1
Elements to Be Included in a Recovery Plan for an Employee Returning to Work

Activity	Tick
Outline the primary objectives, such as abstaining from alcohol during work hours, attending treatment or counselling sessions or improving workplace performance. Give clear guidance on what constitutes success.	
Define how the plan will support the employee in maintaining or returning to their role through the recovery process.	
Include the employee's acknowledgement of the issue and their commitment to actively participate in recovery efforts.	
Allow the employee to contribute to the action plan, ensuring that it is realistic and considers their recovery needs.	
Outline what support the employer will provide, including access to resources (such as EAPs, addiction counselling or rehabilitation services), adjustments and confidentiality measures.	
Engage in an interactive dialogue with employees to determine the accommodations that would be most helpful.	

TABLE 8.1 (*Continued*)
Elements to Be Included in a Recovery Plan for an Employee Returning to Work

Activity	Tick
Detail accommodations, such as adjusted work hours, remote work options or temporary modified duties.	
Recognise the role of family and social networks in an employee's recovery journey and provide resources for employees seeking support from loved ones.	
Establish a schedule for one-on-one meetings with a manager or HR representative to monitor progress and provide ongoing support.	
Set regular intervals for reviewing the employee's progress against the action plan goals, with clear dates for these reviews.	
Clearly state what is expected of the employee, such as attending treatment sessions, meeting performance benchmarks and adhering to workplace policies.	
Identify measurable goals, such as punctuality, attendance levels or meeting job performance standards, that will be used to evaluate progress.	
Reinforce expectations about workplace behaviour, including refraining from alcohol use during work hours and maintaining professionalism.	
Include provisions for modifying the plan if circumstances change, ensuring it remains relevant and supportive.	
Outline how the employee's information will be handled, emphasising confidentiality and limiting disclosure to those with a legitimate need to know.	
Ensure that records related to the action plan are stored securely and separately from general personnel files.	
Define the length of the support period, such as three, six or twelve months, with specific milestones to measure progress.	
Clarify what success looks like, such as completing a rehabilitation programme or maintaining consistent job performance for a defined period.	
Describe how the employee will gradually return to their full duties if accommodations or reduced responsibilities are implemented.	
Assess whether the work environment is supportive and stigma free and workplace culture will not pose a risk to the employee's recovery.	
Before reintegration, conduct a comprehensive assessment to determine the employee's readiness to return to work.	
Ensure the employee has received medical clearance, if necessary, to confirm they are physically and mentally fit to resume their duties.	
Include a plan for transitioning to disciplinary action if the employee does not engage with the action plan, fails to meet agreed-upon goals or poses risks to the workplace.	
Require the action plan to be signed by the employee, their manager and a representative from HR, demonstrating mutual understanding and commitment.	

8.3.4 Transitioning to a Disciplinary Response

Despite the emphasis on support, there are times when switching to disciplinary action becomes necessary. This is a last resort for when an employee repeatedly fails to engage in treatment or continues to pose a risk, despite reasonable accommodations. Before initiating action, you should exhaust all available support options. Follow your established disciplinary procedures, which may include escalations through verbal warnings, written warnings, suspension and potentially to dismissal. Clearly document efforts to assist the employee and check for procedural fairness and compliance with the ACAS Code of Practice to help mitigate the risk of legal challenges.

8.4 RECOVERY OPTIONS

8.4.1 Helplines

All national alcohol helplines operate a 24/7 service and offer free and confidential advice to anyone concerned about their own or someone else's alcohol use. This can be particularly important for those who may feel isolated or embarrassed about seeking assistance in person. The anonymity of helplines can also encourage more people to take the first step towards recovery by discussing their issues without fear of judgement.

They can provide tailored advice on reducing alcohol intake, raising awareness about the potential health impacts, and recognising signs of harmful drinking patterns. Trained advisors offer compassionate, non-judgemental support and information about local resources, including counselling and treatment services. Advisors can help callers to better understand alcohol problems and make informed decisions.

For family members and friends, they also offer support and strategies to navigate the impact of a loved one's drinking on relationships and personal wellbeing. They can provide strategies for talking to someone about their drinking, managing difficult situations and finding help for family members who are indirectly affected.

Helplines bridge the gap between immediate guidance and local treatment options. If you are unsure how to access help for an employee, contacting a national or local helpline is a good first step.

In some areas, there are additional services for special populations, such as over 50s, women or members of the LGBTQ community.

Drinkline is the national alcohol helpline for England and Scotland. In Wales, the helpline is called DAN and covers both alcohol and other drugs as does the Northern Ireland Drug and Alcohol Service helpline. A Google search will bring up the relevant telephone number in each country.

8.4.2 GP Consultation

GPs are often the first port of call for someone with an AUD. They are able to screen for alcohol use and run tests to determine whether alcohol has caused any physical harms, such as blood screen for damage to the liver (liver function tests—see later in this chapter). Knowing their patient's history, they can also advise whether referral to an alcohol treatment service is appropriate or whether general counselling or medication would be more suitable. Regular GP check-ups can help to keep the patient on track and tackle any health issues before they escalate.

However, employees may be unwilling to contact their GP out of concern about what information may be included in their medical records and how this may affect future employment or other aspects of their life. These potential downsides need to be weighed against the benefits of early treatment for alcohol-related health consequences, as sharing this information allows the GP to coordinate care more effectively, taking alcohol use into account when recommending medications or treatments for other health issues. For instance, certain medications can interact negatively with alcohol, and being aware of an AUD helps GPs make safer, more informed prescribing decisions. If your employee is still unwilling to disclose their difficulties to the GP, their choice should be respected, and there are plenty of other options available.

Ultimately though, a supportive GP offers a safe, confidential space to discuss the impact of alcohol use, and they have the expertise to assess any serious health risks, such as liver damage, hypertension or mental health issues, which often accompany prolonged alcohol misuse.

8.4.3 Alcohol Coaching and Sober Coaching

This relatively new intervention is often more flexible than traditional treatment.[2] Coaching sessions can be held online and outside of traditional office hours, making them more accessible for employees. This is a particularly useful option for:

- People whose difficulties do not indicate an entrenched alcohol problem
- 'High risk' drinking scoring 16–19 on the **AUDIT** screening tool
- Anyone who would benefit from short-term support to regain control over their drinking

Coaching provides personalised, solution-orientated support, helping clients navigate the complexities of change, build coping strategies and maintain accountability. Coaches work one-on-one with clients, offering guidance on setting realistic and achievable goals, such as gradually reducing consumption or identifying and addressing triggers. Unlike conventional therapy, which can focus more broadly on mental health, alcohol coaching is specifically geared towards understanding drinking behaviours, modifying habits and supporting behaviour change. Coaches tailor their approaches to everyone's unique goals, needs and progress, which can be highly empowering. They may use techniques like motivational interviewing, cognitive-behavioural strategies and practical, goal-focused exercises, helping individuals learn to resist cravings and manage stress without alcohol so that they learn to replace drinking behaviours with healthier habits.

Accountability is also a core element of alcohol coaching. Regular sessions create a structured check-in process, encouraging clients to reflect on their achievements and challenges. This consistency can be particularly beneficial for those who need ongoing support or who have tried and struggled to reduce their alcohol use on their own. Coaching should create a supportive, judgement-free space, which allows clients to address the social and emotional aspects of their alcohol use openly, making it easier to address underlying factors.

The quality of the therapeutic relationship between coach (or counsellor) and their client is a key indicator of successful treatment, so coaching in this way enables your employees to explore different coaching options and styles to find a practitioner that they are comfortable with.

A note of caution, though: this sector is still unregulated, and anyone can set themselves up as an alcohol/sober coach with no qualifications or experience. There is an excellent resource called Sober Code, a voluntary code of ethics to which coaches can sign up. It also provides a guide to questions that potential clients should ask of coaches before signing up to their services. Coaches are not able to provide formal treatment options such as counselling and cannot make diagnoses or prescribe medication.

8.4.4 Self-Help

In the UK, several self-help options are available for individuals managing AUD, ranging from digital tools to guided exercises and educational resources. These resources provide accessible, flexible ways for individuals to work towards reducing or stopping alcohol use, with the added benefit of privacy and autonomy. Usually based on psychological principles and using the same modalities as traditional treatment, recovery manuals and online programmes take readers through a set of steps to achieve their aims from preparing mentally to maintaining momentum once initial changes have been made. These have the advantage of being confidential and allowing the user to work through modules at their own pace. They can, of course, also be accessed day and night, so help is at hand when alcohol cravings surface or the user is struggling with motivation.

Digital and mobile apps offer practical tools to help users monitor their drinking, set targets and stay on track with their goals. These apps typically provide features such as daily alerts, progress updates and sometimes, educational content to support users in understanding and adjusting their drinking habits. They deliver real-time support and can be used privately, making them ideal for those who prefer a discreet way to manage their alcohol consumption or need flexibility in their approach due to busy schedules.

Online resources and self-assessment tools provide information about the effects of alcohol on health, self-assessment quizzes and tips for cutting back. Self-assessment can give employees an initial understanding of their alcohol use, guiding them in deciding whether professional support might be necessary.

Self-help books, such as those on mindful drinking or CBT-based strategies for managing alcohol use, provide structured programmes that individuals can follow at their own pace. These techniques can help in managing cravings, understanding triggers and developing healthier coping mechanisms for stress. Guided mindfulness exercises are often also available in apps or online and some self-help CBT resources are available through local libraries or NHS digital libraries.

Campaigns such as Dry January and Sober Spring also belong under self-help. Alcohol Change UK, the charity that runs these initiatives, offers additional online support to help participants through sobriety challenges. These are a great way to kickstart changes by providing a time-limited period to pause and reflect on their drinking habits without the pressure to make a permanent commitment. These short (one to three months) abstinence challenges create an opportunity for participants to examine their relationship with alcohol, often revealing underlying patterns that might otherwise go unnoticed.

Participating in the Dry January campaign also has physical and mental health benefits. Research suggests that even short-term abstinence can lead to improved sleep, better concentration, higher energy levels and reduced anxiety and depression symptoms, which often helps individuals realise the impact of alcohol on their well-being. Some studies have shown that people who complete a Dry January challenge tend to drink less for months afterwards,[3] highlighting its potential as an effective, low-pressure starting point for long-term reduction in alcohol use.

Additionally, the campaign's large-scale social aspect provides a sense of shared experience, making it easier for participants to stay motivated and avoid feeling isolated in their journey. The campaign's popularity has led to a broad support network that includes resources such as online communities, discussion forums and social media groups. This affords participants the opportunity to share their experiences, gain encouragement and receive tips from others on the same journey. In the process, they develop healthier coping mechanisms and social strategies, which are helpful in maintaining reduced alcohol consumption and managing triggers even after the campaign ends.

A word of caution: Dry January and other abstinence campaigns are not to be undertaken by any individual who has or is at risk of alcohol dependence. If cessation of drinking results in any physical symptoms of withdrawal, immediate medical advice should be sought (see Chapter 3).

By combining several of these self-help strategies, individuals have a broad toolkit for addressing AUD. These options allow for a versatile, self-directed approach to recovery, which can be empowering and sustainable. When paired with occasional support from healthcare professionals, self-help resources can form a well-balanced and accessible recovery plan. The quality of these tools can vary though, so it's wise to check the credentials of the authors or the programme and any reviews. Accreditation by a governing body or endorsement by the NHS is usually a good indicator.

8.4.5 Mutual Aid

Mutual aid groups, including Alcoholics Anonymous (AA) and SMART Recovery, are a well-regarded feature of the recovery landscape and offer a non-judgemental environment of communal support. In these groups, members connect with others who have experienced similar struggles, helping them feel less isolated and more empowered in their recovery. Regular meetings provide a safe space for individuals to openly discuss their challenges, successes and setbacks without fear of judgement, which can be a powerful motivator and source of encouragement.

Groups like AA use specific recovery models that guide members through a series of steps aimed at improving self-awareness, accountability and behaviour change. AA's well-known twelve-step programme, for example, combines principles of personal responsibility with spiritual reflection to help members address the root causes of their alcohol use. In contrast, SMART Recovery is based on evidence-backed techniques like cognitive-behavioural therapy (CBT) and focuses on helping participants build coping skills and modify unhelpful thought patterns. These techniques are designed not only to reduce alcohol consumption but also to support long-term recovery by equipping members with practical skills that they can apply in real-life situations.

Furthermore, mutual aid groups offer consistent, accessible and often no-cost resources that can complement other treatments. Participants can attend meetings as often as needed, and the lifelong nature of many mutual-aid organisations encourages a sustained commitment to recovery. Many people find that the ongoing support and accountability provided by these groups are instrumental in maintaining sobriety and preventing relapse. That said, they are specifically abstinence-focused, and for some individuals, this will not be their aim, at least initially.

8.4.6 Alcoholics Anonymous

Founded in 1935, AA is a peer-led group that provides a safe environment for members to share their experiences and support one another in their recovery journey. The charity's primary goal is to help its members achieve abstinence and maintain it through mutual support and shared experiences.

One of the central features of AA is its twelve-step programme, which exhorts members to acknowledge their struggles with alcohol and take personal responsibility for their recovery. The steps include admitting powerlessness over alcohol, seeking guidance from a higher power and making amends for past wrongs. This spiritual component, while not mandatory, often resonates with many participants and helps them find a deeper sense of purpose in their recovery efforts.

Meetings are conducted regularly in various public settings such as community centres and churches, making them accessible to a broad audience. The meetings provide a platform for members to discuss their challenges, celebrate milestones and offer support to one another. This community aspect can be a lifeline for those who may feel isolated due to their alcohol use. The life-long nature of AA participation can be challenging for some, as can the insistence on abstinence and the references to a higher power. These difficulties aside, AA has supported millions of users to achieve sobriety worldwide and is the most well-known of the mutual aid societies.

AA is also characterised by its emphasis on confidentiality and anonymity, allowing members to share openly without fear of judgement or stigma. This can create a more comfortable atmosphere for individuals to explore their feelings about alcohol and seek help. As each group is run autonomously, each is unique and an individual who may not have found one group particularly helpful may benefit from trying a different group.

In recent years, AA has adapted its approach to cater to diverse needs, offering various formats of meetings, including face-to-face, online and telephone support. This flexibility allows for a format that suits each individual best, particularly during challenging times like the COVID-19 pandemic, when many people turned to virtual support.

8.4.7 SMART Recovery

SMART Recovery is a self-help initiative that features a network of mutual support groups focused on promoting self-empowerment using evidence-based strategies.

The acronym stands for 'Self-Management and Recovery Training' and it offers an alternative approach to traditional twelve-step programmes such as AA.

A key aspect of SMART Recovery is its emphasis on teaching practical skills to manage addiction effectively. The programme is structured around a four-point system designed to help individuals achieve and sustain recovery. The four points are: *enhancing and sustaining motivation*; *coping with cravings*; *managing emotions and thoughts* and *creating a balanced life*. SMART Recovery encourages members to cultivate a fulfilling lifestyle that will support their recovery journey, including nurturing healthy relationships and engaging in constructive activities.

SMART Recovery sessions are led by trained peer-facilitators who encourage discussions and guide activities. This peer-led approach promotes a supportive atmosphere where participants can openly share their experiences. Unlike some recovery programmes, SMART Recovery does not necessitate complete abstinence to join; it welcomes individuals at any stage of their recovery process.

The programme also provides an array of online resources, such as meetings, forums and educational content, making it accessible for anyone who is unable to attend face-to-face gatherings. This adaptability is particularly beneficial for individuals who are employed or those who might feel uneasy in conventional group settings.

8.5 ALCOHOL TREATMENT

Each local authority funds the provision of alcohol treatment services. These are free at the point of contact and can include a range of different approaches and modalities. In some areas, these services will be jointly commissioned with drug treatment services, but in others will be independent. There are diverse treatments available, depending on the severity of the problem and the aims of the client, ranging from very brief interventions to inpatient treatment and longer-term support.

Alcohol Identification and Brief Advice (IBA), as described in Chapter 4, is routinely delivered in primary care. It's a preventative measure, aiming to identify excessive drinking at an early stage, when it is still relatively easy for the individual to make changes. This light touch approach is the most widely used response to alcohol problems and statistically one in eight recipients will reduce their drinking.[4]

> *IBA involves giving simple advice to people drinking at increasing or higher risk levels who may not be consulting for an alcohol related problem.*
>
> (*British Medical Journal (2010) 340:c1595*)

Extended brief interventions are also carried out in community settings. Typically, these take place over four to six weeks and offer advice offer support with self-monitoring of alcohol consumption, coping with high-risk situations, goal-setting and healthier lifestyle options. Support can generally be accessed in person in community treatment settings or online.

8.5.1 Community Alcohol Treatment

Beyond these shorter interventions, community alcohol treatment services in the UK are designed to provide accessible support for anyone struggling with AUDs. These services can be abstinence harm reduction focused, depending on patient goals.

Community-based programmes typically offer a range of services, including talking therapies (particularly CBT or Motivational Interviewing modalities), group therapy and peer support sessions. They may also provide medical interventions, such as detoxification or medication-assisted treatment, tailored to individual needs.

One of the key components of community services is an emphasis on aftercare and ongoing support. Many programmes provide relapse prevention sessions, ensuring that clients have the resources they need even after completing formal treatment. These services often work closely with local health authorities and charities to guarantee a holistic approach to alcohol treatment. For example, concurrent support for mental health challenges is often beneficial for clients in treatment.

The availability of community alcohol treatment varies by region, with services accessible through local councils, healthcare providers and charities. Individuals seeking help may be able to self-refer or may require referral by healthcare professionals.

8.5.2 Emergency Treatment

Emergency treatment encompasses situations where an individual is experiencing severe alcohol-related complications, such as alcohol poisoning, withdrawal symptoms or other medical emergencies associated with heavy drinking. The primary goal is to stabilise the patient and prevent life-threatening conditions.

In cases of acute alcohol intoxication, symptoms may include confusion, vomiting, seizures, slow or irregular breathing and hypothermia. Emergency medical services will typically perform an assessment and may provide interventions such as oxygen therapy, intravenous fluids or medications to protect the airway and prevent aspiration. In severe cases, hospitalisation may be required to monitor vital signs and manage complications, including the risk of respiratory failure or coma.

For individuals experiencing withdrawal symptoms, which can range from anxiety and tremors to severe complications like delirium tremens (DTs), medical professionals will conduct thorough assessments to determine the best course of action for safe detoxification. Treatment is likely to involve the administration of intravenous thiamine and a course of benzodiazepines to ease withdrawal symptoms and prevent seizures.

8.5.3 Structured Day Programmes

These programmes typically offer a combination of therapy, education and support services within a structured environment, allowing participants to engage in treatment while still maintaining some aspects of their daily lives, such as work or family commitments.

Clients often attend a centre for one or more days a week, where they can engage in various therapeutic activities, including individual counselling, group therapy and skill-building workshops. Day programmes help clients develop regular routines, explore positive coping strategies, understand their triggers and establish a support network with peers who are facing similar challenges.

Many structured day programmes also incorporate educational components to raise awareness about the effects of alcohol, the importance of a healthy lifestyle and relapse prevention techniques. This holistic approach can improve treatment outcomes, as it addresses not just the physical aspects of alcohol dependence but also the psychological and social factors involved. Research indicates that individuals who participate in structured day programmes often report increased motivation and a better understanding of their addiction, leading to more effective recovery strategies.[5] These programmes can be especially beneficial for those who may not require intensive inpatient treatment but still need regular support and guidance to overcome their alcohol-related issues.

8.5.4 Community Detoxification

Community alcohol detoxification (detox) offers individuals a safe way to reduce or stop alcohol use outside of a hospital, typically in their own home or a local health facility. This programme is aimed at people with mild to moderate alcohol dependence who want to avoid the expense and life disruptions that can come with inpatient detox, but who still are likely to experience physical symptoms when they stop drinking and require medication and monitoring to manage withdrawal symptoms safely.

The process usually begins with a thorough assessment by a healthcare professional to consider the individual's physical and mental health, drinking history and the potential for severe withdrawal symptoms. Based on this assessment, a personalised plan is developed to guide the detox process. This plan may involve gradually reducing alcohol consumption or, if necessary, managing withdrawal with specific medications.

While community detox is less intensive than inpatient care, it still provides access to medical oversight. Participants may have regular check-ins with a nurse or keyworker, either by phone or in person, to monitor their progress and respond to any arising issues. In some cases, medications, such as benzodiazepines and thiamine, are prescribed to help manage the discomfort of withdrawal and reduce the risk of serious symptoms.

Participants are encouraged to involve supportive family members or friends in their journey, for additional motivation and a sense of accountability.

Community alcohol detox also focuses on symptom management, addressing common withdrawal effects like anxiety, insomnia and nausea. Self-care practices such as hydration, good nutrition and strategies to handle cravings are emphasised. If withdrawal symptoms become too intense or there are complications, the healthcare team may recommend a shift to inpatient care for the individual's safety.

Aftercare is an essential part of a successful detox, as it helps prevent relapse and supports long-term recovery. Many community detoxes include follow-up planning, encouraging individuals to continue with counselling, join support groups or enter outpatient addiction treatment. Developing healthy coping skills to replace drinking and finding positive outlets for stress and emotions are central to staying sober.

8.5.5 Residential Rehabilitation

These interventions typically last from a few weeks to several months and are designed to address the physical, psychological and social aspects of addiction.

Residential rehabilitation centres offer comprehensive treatment plans that often include medical detoxification, individual and group therapy sessions, educational workshops and holistic therapies. The goal is to help individuals understand their relationship with alcohol, develop coping strategies and build a support network. Rehab centres often also address underlying mental health issues, such as anxiety or depression, that can contribute to alcohol dependency.

The facilities vary in terms of their approach and treatment methodologies. Some may use evidence-based psychotherapeutic practices or the twelve-step model, while others may integrate alternative therapies such as mindfulness, yoga or art therapy to enhance the recovery process. Aftercare support is an important component, as it helps clients transition back into their daily lives and maintain their sobriety post-rehabilitation.

Costs for residential rehab can vary tremendously, with options available through the National Health Service (NHS) for those who qualify, as well as private facilities that may require out-of-pocket payment or insurance coverage. Access to these services can be facilitated through referrals from healthcare providers, addiction services or self-referral.

8.6 TALKING THERAPIES

8.6.1 Cognitive Behavioural Approaches

Cognitive behavioural therapy (CBT) is present-focused and stresses the development of practical skills to manage cravings and recognise triggers to drink. Techniques may include problem-solving strategies, goal-setting and behavioural experiments that encourage clients to engage in activities that do not involve alcohol. Research has shown that **CBT** can lead to significant reductions in drinking and improvements in related psychosocial issues, such as mood disorders.[6]

Therapists work with clients to identify triggers for drinking, such as stress or social situations and explore the underlying beliefs and thoughts that may contribute to these triggers. By addressing unhelpful thinking patterns—such as self-blaming or catastrophising—clients learn to challenge their automatic thoughts about alcohol and replace them with more balanced perspectives.

Developed from **CBT**, Motivational Enhancement Therapy and Dialectical Behaviour Therapy are often offered alongside traditional **CBT**. Similar in approach, they focus on building readiness to commit to treatment and managing intense emotions and impulses respectively. Generally, **CBT** is delivered in individual sessions or as part of group therapy programmes. It can be integrated with other treatment modalities, including motivational interviewing, contingency management and pharmacological treatments, to provide comprehensive care. Studies suggest that **CBT** is effective for both mild and severe AUDs, making it a versatile option in various treatment settings.

8.6.2 Motivational Interviewing

Motivational interviewing (MI) is a client-centred, directive approach aimed at enhancing an individual's motivation to change. This approach emphasises the exploration of the client's ambivalence towards their alcohol use using open-ended questions, reflective listening and affirmations to facilitate a conversation that helps clients articulate their thoughts and feelings about drinking. The goal is to empower clients to identify discrepancies between their current behaviour and their broader life goals, rather than trying to persuade or coerce changes.

Studies indicate that MI can prompt reductions in alcohol use and enhance engagement in treatment. The approach is particularly effective in initial sessions, where building rapport and maintaining motivation are paramount. Further information about treatment options can be found on the NHS or Alcohol Change UK websites.

8.6.3 Medication-Assisted Treatment

Several medications are available on prescription that may help with different aspects of alcohol dependence. Those licensed for use in the UK are listed next. An employee who is prescribed any of these medications should consider the impact on their ability to perform their normal work duties. It is sensible to require employees (e.g. via your alcohol or substance use policy) to disclose any medications that may cause impairment to HR.

It's also imperative that medications are not viewed as a 'cure'. Whilst they can impact alcohol consumption and relapse potential, they work most effectively alongside psychological therapies that address the root causes of the addiction.

TABLE 8.2
Table of Medications Used in Medication-Assisted AUD Recovery

Medication	What is it?
Chlordiazepoxide	Brand name *Librium*. A benzodiazepine used to treat anxiety, reduce muscle spasms and ease insomnia. Helpful for alcohol detoxification, it reduces withdrawal symptoms. Used as a short-term treatment only as it can lead to dependence. Can cause drowsiness.
Naltrexone	Diminishes the 'reward' experienced when drinking and can be used in two ways: either to decrease cravings for alcohol and encourage abstinence or to support a moderation goal by reducing cravings.
Disulfiram	Brand name *Antabuse*. Interferes with the way the body metabolises alcohol. Used once the individual has achieved abstinence. If someone taking this medicine drinks alcohol, it quickly causes a severe, unpleasant and potentially dangerous reaction and knowledge of this can help to prevent a return to drinking.
Acamprosate	Brand name *Campral*. Used as soon as the patient stops drinking to maintain abstinence by reducing cravings and can be prescribed for up to six months. It can be paired with other medication treatments, such as naltrexone.
Nalmefene	Brand name *Selincro*. Can be used for a controlled drinking goal. Recommended for high-risk drinkers. Should always be used alongside talking therapies. Nalmefene reduces cravings and should be taken in anticipation of drinking about one to two hours before drinking rather than every day.

Recent studies are showing positive results for psychedelic therapy as a treatment for alcohol dependence. Psychoactive drugs such as MDMA (ecstasy) and psilocybin (magic mushrooms) have been effectively used to treat alcohol disorders but are not currently licenced in the UK for this purpose.

8.7 FAMILY AND SOCIAL NETWORKS AS RECOVERY CAPITAL

Any support for a dependent drinker's recovery should not come at the expense of the wellbeing of the supporter. Caring for someone through alcohol dependence and recovery is hard. It will take a toll on the loved one's resilience and often their own physical and emotional health and social support will suffer too, and it is sometimes necessary for family members to distance themselves from their loved one. In some cases, a dependent drinker's behaviours will be aggressive or abusive and this can have a wholly destructive impact on family life.

However, where family and social networks *are* able to offer support, they can be instrumental in an individual's recovery. Reassurance, encouragement and understanding all help ease the emotional strain that often accompanies treatment. This emotional backing is particularly important during moments of frustration or temptation, as it can give individuals the strength to persevere and stay committed to their goals.

In addition to emotional support, family and social networks help with accountability. Those closest to the individual can assist in monitoring progress, helping the person stay true to their recovery objectives. This accountability, when handled positively, can strengthen the individual's resolve and help them to stay committed to their long-term goals. Loved ones can also play a role in identifying any setbacks or struggles, providing timely support and intervention when necessary.

Lastly, positive reinforcement from family and friends can boost motivation. Encouragement and praise for progress, no matter how small, help reinforce positive behaviours This not only strengthens the individual's resolve but can also help them navigate challenges in the workplace, where maintaining a healthy balance and professional success are often necessary for long-term recovery.

8.8 FAMILY THERAPY AND INVOLVEMENT

Education is of huge benefit in building an effective support system. When family members and friends understand the challenges of alcohol dependence, they are better equipped to offer the appropriate support to their loved one and take care of themselves. Educating loved ones about the nature of addiction, withdrawal and recovery can help them approach the situation with empathy and patience, avoiding judgement or frustration. This understanding can also lead to better communication, allowing individuals to feel more comfortable opening up about their struggles and progress without fear of misunderstanding.

Family therapy involves family members in the recovery process, addressing relationship issues and improving communication. Some modalities, such as the *5-Step approach*, aim solely to support the family members to meet their own needs and reduce the negative impact of their loved one's drinking on their daily lives. This is a client-centred model that provides information, teaches improved coping techniques and helps the client to identify additional sources of support.

Other models, such as *Community Reinforcement and Family Training (CRAFT)*, have the twin aims of supporting the family member and encouraging the drinker to access recovery options.

Non-alcohol use-specific therapies, such as Brief Solution Focused Therapy and Behavioural Couples Counselling, are also beneficial in supporting loved ones to manage the impact of someone's drinking. These may be available to employees through your EAP.

Further information about these options can be found on Adfam's website.

8.9 PROS AND CONS OF RECOVERY OPTIONS

TABLE 8.3
Pros and Cons of Different Recovery Options Available in the UK

Option	Pros	Cons
Helplines	Free and confidential to access, a useful first step to access information and advice	Don't provide ongoing or tailored support. Cannot assess the nature and extent of the problem
Coaching	Supportive, goal-focused intervention, often useful for evening/weekend access and online options	Unregulated so coaches may lack qualifications or experience. Can be expensive
Self-help	Flexible to access and confidential. Can be effective developing coping skills	No external accountability and only suitable for high-risk drinking, rather than dependence
Mutual aid	Supportive peer-led environment with strong accountability, often available over longer term	Unregulated and lacks clinical input that may be required in response to severe dependence
Talking therapies	Can help individuals address the emotional and psychological aspects of their alcohol problems alongside skill development	There may be waiting times or restrictions on appointment availability, often time limited. Private therapy can be costly
Structured day programmes	Typically combine therapy, support, education and skill development whilst establishing a consistent routine	Often prohibitively time consuming for those in employment
Residential treatment	Intensive treatment offering the benefits of day programmes but often with additional medical or psychiatric input for those with more complex needs. Can be coupled with medical detox	Funded places are difficult to access and may be located at a great distance from the employee's home. Private treatment can be financially prohibitive. Employee will need to be absent from work for weeks or months
EAP	Can be accessed for alcohol problems or for loved ones supporting a dependent drinker. Can explore other work-based difficulties	May not provide specialist addiction treatment. Employees may be reluctant to access the service due to stigma or privacy concerns

QUESTION FOR REFLECTION

If you were in a position to benefit from alcohol treatment, which option would you choose and why?

NOTES

1 Vitality Health Insurance, Britain's Healthiest Workplace (https://www.vitality.co.uk/media/britains-healthiest-workplace-alcohol/).
2 Sober coaching is a sub-category of alcohol coaching that focuses on supporting clients to achieve total abstinence.
3 De Visser, R. O., 2019. Evaluation of Dry January 2019; de Visser, et al., 2017. The Growth of 'Dry January' Promoting Participation and the Benefits of Participation.
4 Raistrick, D., Heather, N. and Godfrey, C., 2006. Review of the Effectiveness of Treatment for Alcohol Problems. *National Treatment Agency for Substance Misuse* (https://www.drugsandalcohol.ie/6153/1/3246-3419.pdf).
5 Song, et al., 2022. Effectiveness of Therapeutic Community Program on Resilience and Change in Lifestyle in People With Alcohol Use Disorder. *Journal of Addictions Nursing* (https://pubmed.ncbi.nlm.nih.gov/37140413/).
6 Magill, et al., 2019. A Meta-Analysis of Cognitive-Behavioural Therapy for Alcohol or Other Drug Use Disorders: Treatment Efficacy by Contrast Condition. *Journal of Consulting and Clinical Psychology* (https://pmc.ncbi.nlm.nih.gov/articles/PMC6856400/).

9 Creating an Alcohol-Aware Workplace

9.1 INTRODUCTION

Hopefully, by now you've developed an alcohol policy and evaluated how best to respond to the risks of alcohol at work. You may already have considered what your colleagues and employees need to know, understand and be able to make sure that alcohol isn't a threat to their wellbeing. This is the pay-off; this is where you can start to forge a sustainable plan for the future. Building an alcohol-aware workplace can not only improve wellbeing but also lead to valuable resource savings. In this chapter, we will explore good practices for awareness campaigns, education and training programmes and promoting an environment that places a premium on health.

QUESTION FOR REFLECTION

If you were to put together a working group to tackle potential harm from alcohol and improve wellbeing in your workplace, who would you want to involve and why?

9.1.1 The Evidence Base

The benefits of integrating alcohol awareness into your wellbeing strategy are underpinned by a large body of evidence.[1] Conversely, a workplace that normalises or encourages excessive alcohol use, such as through frequent social events involving alcohol, can alienate employees who abstain while perpetuating unhealthy drinking patterns among others.[2] Whatever the starting point of your workplace alcohol culture, embedding alcohol awareness into your workplace wellbeing activities offers a proactive approach to mitigating these challenges. Early intervention programmes such as educational campaigns can help employees recognise the risks associated with excessive drinking and identify early signs of dependence. This can lead to timely support and reduce the likelihood of long-term health and workplace consequences. Coupling alcohol awareness with broader initiatives, including mental health promotion, sleep hygiene and stress management, ensures a holistic approach that aligns with evidence-based best practice.[3]

Research demonstrates that incorporating alcohol awareness initiatives provides tangible benefits. For example, a 2018 study by the Society for Human Resource Management[4] found that companies with proactive substance misuse programmes experienced reduced absenteeism rates, improved employee engagement and

DOI: 10.1201/9781032648927-10

enhanced workplace morale. When these are combined with a clear, relevant policy, there's a shift towards healthier behaviours and attitudes, as well as a more positive and inclusive workplace.[5]

QUESTIONS FOR REFLECTION

What is the primary purpose of your alcohol awareness initiatives? Are you aiming to reduce alcohol-related incidents, increase awareness, provide support, all of the aforementioned or something else?

Are you focusing on all employees, specific departments or individuals with identified risks?

What will your key performance indicators be?

Do you have a timescale for the completion of activities, or will your programme be ongoing?

9.2 PREPARING YOUR PROGRAMME

9.2.1 The Cultural Audit

The rationale for conducting a cultural audit has already been outlined in a previous chapter* but it bears repeating that you will need to consult your staff if you're going to identify the specific problems that you're trying to resolve and the risks that your employees face.

A well-planned audit can yield numerous benefits for both employees and the organisation. One salient outcome is heightened awareness among employees about the organisation's stance on alcohol consumption and its commitment to a healthy work environment. Increased awareness helps to harmonise employees' behaviours and expectations with the company's values and wellbeing initiatives. This can help reduce the stigma often associated with seeking help for alcohol-related challenges, making it easier for employees to access the tools they need. Employees who feel supported in this way are more likely to experience a positive work environment, contributing to better morale and engagement across the business.

The audit process also creates opportunities for open communication between employees and management, encouraging candid discussions about alcohol-related concerns. By establishing a dialogue, employees may feel more empowered to share their experiences or seek support. Furthermore, the audit helps identify potential risks or issues early, so that you can take proactive measures to mitigate them, tackling problems before they escalate.

To maximise the effectiveness of your audit certain good practices should be incorporated. First, inclusivity and diversity must be considered throughout the process. The audit should be designed to reflect the unique perspectives and needs of your diverse workforce, taking into account factors such as gender, age, cultural backgrounds and religious beliefs. For instance, certain cultural or religious groups may have specific attitudes towards alcohol that need to be respected and understood in the context of workplace policies.

As always, you'll need to establish methods for ensuring confidentiality and privacy. Employees must feel assured that their participation and responses will remain anonymous and that their feedback will be used solely to improve the workplace environment. A lack of trust in this regard can deter participation and undermine your goals. Make sure you are clear in your communications about the purpose and protections associated with the audit.

9.2.2 Employee Survey

A good place to start is with an employee survey. A well-designed survey is helpful for gathering information on employees' perceptions, experiences and attitudes regarding alcohol in the workplace. It also raises awareness of your commitment to tackling alcohol harm and can open up a conversation about the employer's role.

KEY SURVEY QUESTIONS AND THEIR RATIONALE

Do you believe the organisation has clear policies regarding alcohol use in the workplace?
This question assesses the perceived clarity of existing policies and whether employees are aware of them.

Do you have any concerns about a colleague's drinking?
This question helps identify whether alcohol-related behaviours are occurring and if there are specific incidents of concern.

Do you feel comfortable discussing alcohol-related concerns with your supervisor, Occupational Health or Human Resources departments?
This gauges the openness of communication routes regarding alcohol-related issues.

Are you aware of any employees who have sought help or support for alcohol-related problems?
This question explores whether employees are aware of available support systems and whether they are being utilised.

Do you think there is a social expectation to consume alcohol during work-related events or celebrations?
This starts to examine social pressures and expectations surrounding alcohol use in the workplace.

Are there comparable non-alcoholic options available during work-related events where alcohol is served?
This question evaluates whether the organisation provides inclusive options for employees who choose not to consume alcohol.

Have you experienced any discomfort or concerns related to alcohol use by colleagues during work-related events?
It assesses whether employees have experienced discomfort or discomfort due to alcohol-related behaviour.

Does the organisation provide sufficient information and support to tackle alcohol harm?
This explores whether employees believe that the organisation's alcohol response is adequate.

Have you received any alcohol-related training or education from the organisation?
This question determines the extent to which employees have received information or training regarding alcohol awareness and policy.

How would you rate the organisation's approach to alcohol-related issues in the workplace?
This question gathers employees' general impressions, allowing for an assessment of the organisation's current culture.

Distribute your survey to all employees, ensuring anonymity and confidentiality to encourage honest responses. You can use online survey platforms, paper surveys or a hybrid, depending on what works best for your organisation. Keep the window for responding short—no longer than a week or two, to encourage respondents to act quickly. When you have your data, you can start to analyse the responses to identify trends, patterns and areas of concern. Consider demographics (e.g. department, role, tenure) when analysing data to distinguish potential variations in responses. For example, office-based staff may have very different level of awareness of your policy and its implications when compared to employees in safety-critical roles.

If the survey raises further questions (*why do people think we aren't doing enough? How come they don't remember the training event we held last year?*), consider conducting focus group discussions or individual interviews with a sample of employees to gain deeper insights into their experiences and perceptions.

You will now be in an excellent position to evaluate the findings and identify any action points, which might include increasing communication about your policy content, offering alcohol awareness information or enhancing support systems for employees with alcohol-related concerns. Share your findings with employees and communicate any changes or actions that will be taken as a result. To facilitate transparency and involve employees in the process of implementing changes, you could consider setting up an Employee Resource Group to carry the work forward.

9.2.3 ERGs

Employee Resource Groups (ERGs) are employee-led groups that promote inclusion, support and education on specific topics, making them well-suited for addressing sensitive issues like alcohol awareness. Establishing an ERG focused on alcohol provides a confidential and supportive environment where employees can discuss concerns, share experiences and seek advice without fear of judgement. Many employees may hesitate to approach managers or HR about personal struggles related to alcohol due to stigma or fear of repercussions. An ERG led by peers helps normalise these conversations, making it easier for individuals to seek help. They can also enhance the reach and impact of your awareness-raising initiatives by helping to develop tailored programmes that resonate with the workforce. The group can organise events, workshops and campaigns that address alcohol misuse in relatable and accessible ways. Because ERGs are typically run by employees, they can bring grassroots insights into what resonates best within the company's values, ensuring initiatives feel authentic and engaging.

Your ERGs can work with leadership and HR to synchronise alcohol awareness initiatives with other health programmes, such as mental health resources, stress management training or substance misuse support services. By reporting on the group's activities and successes, the ERG showcases the tangible impact of its efforts. Leaders who actively support the ERG by participating in events or sharing personal reflections send a powerful message about the organisation's commitment and ethos. You may also find that employees are more likely to engage with and trust initiatives led by their peers. ERG members can serve as advocates and champions of alcohol awareness, bridging the gap between employees and resources.

Workforces are diverse, and alcohol-related issues may impact employees differently depending on cultural, social or personal factors. An ERG can create a platform to address these nuances, ensuring that resources and programmes are inclusive and representative of all employees' needs. For example, the group might organise discussions that explore cultural attitudes towards alcohol or support initiatives for employees exploring sobriety.

9.2.4 Resource Allocation

Once you've established your aims and devised a strategy for implementation, you'll need to allocate resources across several key areas. Begin by determining the budget necessary to support these initiatives, accounting for expenses such as creating educational materials, organising workshops and accessing external expertise such as counsellors, trainers or motivational speakers. Personnel allocation is another consideration, requiring the involvement of wellbeing coordinators, HR professionals, ERGs or external experts to manage and deliver the initiatives.

Resources should also be allocated for technology and tools, such as online platforms for training, communication and data collection. You could also invest in training and development so that employees and leaders involved in the initiative are confident that they have skills and knowledge needed to carry out their roles successfully. It's wise to also allocate resources for monitoring and evaluation so that you can assess the impact of the initiatives. You may need data analysis software, for example and personnel to input data. To maintain relevance, establish mechanisms for collecting employee feedback and allocate resources to analyse this input and implement necessary changes. Continuous improvement should be a focus, ensuring resources are available to refine initiatives based on evolving needs and insights.

Communication methods also require support, including resources for intranet updates, newsletters and workshops to keep employees informed. Consider partnerships with external organisations where appropriate and allocate funding to support these collaborations. Make resources available for immediate response in case of crises or emergencies and develop contingency plans for addressing unforeseen challenges that may arise.

9.2.5 External Consultants

One additional aspect to consider in conducting your audit is the use of external expertise. Depending on the size and complexity of your organisation, it may be beneficial to engage external experts in workplace culture, alcohol or addiction prevention. Consultants can bring several advantages to the process, including credibility. They can offer an objective perspective and unbiased assessment of your company's policies and practices, which can highlight areas for improvement that might be overlooked by internal staff.

External experts also bring specialist knowledge that can greatly enhance the design and implementation of effective alcohol prevention regimes. They stay up to date with the latest research, industry standards and legal requirements so you can be sure that the strategies developed are not only based on proven methods but also compliant with regulatory frameworks. Moreover, engaging experts ensures that policies and initiatives are comprehensive, bespoke and sensitive to the specific needs of the organisation. They can advise on the creation of fair and transparent alcohol-related policies and can also help to navigate different attitudes and responses to your policy to guarantee that all voices are heard.

When considering external expertise, check that the consultants you choose have a solid track record in workplace assessments and substance misuse prevention. Collaborate closely with them to align their expertise with your organisation's unique goals and challenges. An experienced consultant can also play an essential role in assessing the effectiveness of implemented programmes. Their expertise in evaluating workplace health initiatives can help to make sure that alcohol awareness campaigns or support services are achieving the desired outcomes. Through surveys, feedback mechanisms and performance evaluations, external experts can help to identify what is working well and suggest improvements where necessary.

CASE STUDY

A financial services firm has established an annual 'Wellness Week' that features a wide range of activities, including yoga classes, healthy eating seminars and alcohol awareness workshops. As part of the company's commitment to employee health, managers are required to attend at least one wellbeing event, reinforcing the importance of these initiatives throughout the organisation.

9.2.6 The Importance of Awareness and Education

The type of education and awareness activities that you undertake will be determined by your audit and any identified needs. Where there are safety-critical roles, the focus should be on the direct impact of alcohol on performance, safety and potential harm to others.

Service-based sectors, like hospitality and retail, should emphasise customer interactions, responsible service practices and the handling of intoxicated patrons.

Naturally, one of the foremost reasons for implementing alcohol education is to address the potential harms of alcohol misuse. Your policy will set out the rules and penalties for infringement, but your education programme helps employees understand why the rules exist and arms them with the knowledge to recognise and mitigate risks associated with alcohol. This proactive approach can be especially important in industries where safety is a factor.

Armed with data from your audit, your next step will be to set objectives. The process begins with a clear understanding of the primary purpose of the activity. You'll need to determine your goal: do you want to reduce alcohol-related incidents, increase awareness about the risks associated with alcohol misuse, provide resources and support for employees or a combination of these objectives? Clarifying the purpose serves as the foundation upon which all other elements of the initiative are built.

Once the purpose is established, specifically you need to define your objectives or key performance indicators. These metrics provide a way to quantify the impact of the initiative. If you select the right **key performance indicators (KPIs)** you'll be able to guarantee that progress is measurable and that the outcomes can be evaluated objectively. More on this in Chapter 10. I would advise using SMART criteria—specific, measurable, achievable, relevant and time-bound. For instance, a measurable objective could be to *reduce alcohol-related absenteeism by 10% within the next twelve months*. This goal provides a concrete target and allows for effective evaluation of the initiative's success. Objectives should also be realistic, ensuring that they can be achieved within the given resources. It's also helpful to set clear deadlines for achieving objectives, with milestones along the way to monitor progress. These periodic assessments provide opportunities to make adjustments so that the initiative remains on track.

Understanding your target audience is another major factor. Defining whether the initiative will address the needs of all employees, specific departments or those identified as at higher risk (due perhaps to workplace roles or behavioural indicators) ensures that the programme is appropriately tailored. A light touch approach for all employees may raise general awareness, while targeted initiatives can address identified challenges faced by specific groups.

9.2.7 Ongoing Monitoring

Creating an alcohol-aware workplace requires sustained effort. You'll need to embed various practices so that a positive message is both established and maintained. Central to this effort is the recognition that the journey doesn't end with initial implementation. Ongoing monitoring, employee engagement and iterative improvement are essential for long-term success. For instance, follow-up surveys can gauge employees' awareness of available resources or their willingness to participate in alcohol-related wellbeing activities.

A cultural audit, therefore, should not be viewed as a singular event. Rather, it should be recognised as a continuous practice to evaluate and enhance your workplace ethos surrounding alcohol. Regular assessments, including anonymous surveys and focus groups, can provide valuable insights into the evolving perceptions, behaviours and concerns of employees. These efforts help identify new challenges, allowing you to adapt and refine your strategies. Gathering data on reductions in absenteeism, incidents involving alcohol or increased utilisation of counselling services can serve as indicators of success.

Celebrating successes along the way is equally important. When positive changes result from these efforts—whether it's a noticeable reduction in alcohol-related incidents, higher engagement with support services or improved employee morale—organisations should openly recognise these achievements. Celebrations can include public acknowledgement during staff meetings, sharing progress updates in newsletters or even offering small rewards to teams or individuals who have championed wellbeing initiatives. These actions will not only spotlight the great things that are being achieved but also inspire further engagement and buy-in from employees.

To maintain momentum, you will need to establish robust feedback mechanisms. Employees should feel encouraged and empowered to voice their concerns or share suggestions related to alcohol awareness initiatives. Anonymous options such as (digital) suggestion boxes or open discussions facilitated by ERGs or wellbeing champions can create opportunities for honest feedback. By addressing this input in tangible ways, you can demonstrate the organisation's responsiveness to employee issues.

9.2.8 Addressing Common Concerns and Objections

You may have concerns about incorporating alcohol awareness into your wellbeing agenda, but these can be addressed through proactive communication and thoughtful project design. One common concern is the potential stigma or breach of privacy that employees might associate with participation. To combat this, emphasise that all initiatives are confidential and non-judgemental, reinforcing the organisation's commitment to protecting employee privacy. Offering a range of activities, including those that allow for anonymous participation, can help.

Another issue might be the perception that focusing on alcohol-related matters could negatively impact productivity. This concern can be mitigated by highlighting the long-term advantages of having a healthier workforce. Similarly, worries about the cost of these initiatives can be countered by demonstrating the cost-effectiveness of preventive measures compared to the expenses associated with untreated issues, such as absenteeism, reduced efficiency or workplace accidents. Chapter 10 illustrates that by measuring the outcomes of your alcohol initiatives, you can set future budgets, demonstrate the return on investment (ROI) and justify any expenditure.

Resistance from employees is another challenge that can arise, but this can be managed by ensuring that participation is entirely voluntary. Offering a range of options and respecting individual choices helps to create a non-intrusive approach while still showcasing the potential personal and professional benefits of participation. Lastly, employees may question the relevance of alcohol awareness initiatives to their specific workplace context. You can draw attention to the links between alcohol-related issues and factors like workplace performance, safety and wellbeing to help staff see the value and applicability of your alcohol initiatives.

9.3 COMPONENTS OF A GOOD ALCOHOL EDUCATION PROGRAMME

Having conducted a cultural audit and determined your goals, you can now plan what you want to include and how best to impart information and develop skills. Next, you will find some key areas for consideration. Whatever your content, your messages should resonate with employees, reflecting the realities of the workplace. You'll also need to consider any industry-specific challenges and the diversity of your employees' needs. I've found that the use of straightforward language and real-world examples helps to create content that is both engaging and impactful.

9.3.1 What Employees Need to Know

When employees are aware of the consequences of alcohol misuse, they are more likely to adopt healthier habits, which translates into fewer sick days, improved focus and more consistent job performance. Raising awareness about the short- and long-term health consequences of alcohol abuse empowers your workforce to make informed decisions about their wellbeing. Increased awareness has been identified as one of the key drivers of behaviour change.[6]

Next are some of the most requested and effective elements of an awareness campaign. You could incorporate these into a series of webinars or a poster campaign or wellbeing days. It's up to you to prioritise your topics depending on your audit findings. You'll see from these topics that the focus is rarely on dependent drinking but is firmly aimed at the millions of us who tend to drink too much or too often and don't really consider any potential negative consequences.

- *Alcohol and the body* Educating employees on how alcohol affects physical health, including the impact on the brain, liver, heart and immune function. Explain the weekly recommended maximum number of units and how this translates into drinks. This doesn't need to include scare tactics; for most people, the risks of organ damage, cancer and dependence will not resonate, but increased susceptibility to viruses and infections, weight gain and tiredness might.
- *Mental health effects* Exploring the connection between alcohol use and mental health conditions such as anxiety, depression and stress. The impact of alcohol on sleep and the links to low mood, anxiety, memory loss and 'brain fog'. You could also consider alcohol and the menopause as a topic as excessive drinking can exacerbate both physical and psychological symptoms in mature women.

- *Workplace Implications* How alcohol can impair decision-making, reduce productivity and increase safety risks. It's particularly important to include information on drink-driving laws and how to calculate when the body will have processed all units of alcohol consumed. This is just generally useful information for anyone who drinks to have.
- *Legal and ethical considerations* Outline workplace policies on alcohol use and the potential consequences of violations. Sector-specific information about zero-tolerance, workplace testing and other preventative regulations. You can illustrate with practical examples of how the policy is implemented, too.
- *Recognising signs of alcohol misuse* Train employees and managers to recognise the signs of problematic drinking. Cover behavioural, physical and emotional indicators of alcohol misuse, such as those outlined in Chapter 4. This component should stress that identifying these signs is not about judgement but about offering support and intervention when needed. Your policy should give guidance on what help is available and how to access it.
- *Strategies for reducing alcohol consumption* Equip employees with practical strategies to manage their alcohol consumption responsibly. Topics may include setting limits and knowing personal boundaries, learning alternative ways to cope with stress or social pressure and exploring alcohol-free lifestyle options and the benefits of moderation. You can also provide self-assessment tools, such as the AUDIT (see Chapter 4) or direct employees to external resources.
- *Communication and conversations* Encourage open conversations about alcohol use. Training sessions should include techniques for initiating and navigating these discussions, particularly for managers or team leaders who may need to address sensitive issues with employees. Sessions might explore using non-judgemental language and active listening, choosing an appropriate time and setting for these discussions or knowing how and when to escalate concerns to HR, OH or professional resources.

9.3.2 Manager-Focused Training

Training should empower managers to lead by example and support their teams effectively. They will need to know how to interpret and implement your alcohol policy and how to identify potential alcohol-related issues and take appropriate action. As managers, they will already have the skills to actively listen to and talk to their direct reports about any difficulties they may have. What's often missing is the confidence to talk about alcohol. This is where interactive scenarios that simulate real-life situation can be beneficial, helping managers to practice their skills in a safe environment. For example, it may be useful for managers to work through situations such as handling an employee who appears intoxicated at work, discussing alcohol use with a colleague who seems stressed or withdrawn or planning an alcohol-free social event.

9.3.3 Other Considerations

Despite the pervasive presence of alcohol in UK society, where it is often embedded in social traditions, celebrations and workplace networking events, taboos and stigma surrounding AUDs remain deeply entrenched. Many individuals who experience challenges with alcohol consumption face social judgement, making it difficult to seek help or discuss their struggles openly. This stigma is perpetuated by misconceptions about alcohol dependency, which can lead to the view that it is a personal failing rather than a complex issue influenced by biological, psychological and societal factors.

Moreover, cultural norms in the UK that equate drinking with social bonding can exacerbate the problem, as those who struggle with alcohol may feel isolated or pressured to conform. The stigma is further compounded by limited public understanding of the spectrum of alcohol misuse, which ranges from occasional binge drinking to chronic dependency. By nurturing a non-judgemental environment, you can encourage employees to share their concerns without fear of repercussions. Effective leadership is a must-have; when managers openly support these efforts and emphasise the importance of employee wellbeing, it sets a tone of acceptance and understanding.

Although activities and communications around policies, alcohol testing and regulatory requirements should be mandatory, when implementing wellbeing initiatives in the workplace, make it clear that participation is entirely voluntary. Employees should have the autonomy to choose whether or not to engage with these initiatives based on their personal circumstances and preferences. Communication about events and campaigns should emphasise that participation is optional, avoiding any language or practices that could be interpreted as coercive or as creating an obligation. Employees should face neither stigma nor judgement whether they opt in or out. If you can provide confidential ways to engage, such as private consultations or anonymous surveys, this can help maintain comfort and privacy. Don't be afraid to request regular feedback from employees on how the initiatives are designed and presented; this can help to ensure they remain considerate of everyone's comfort levels. Ideally, your alcohol campaign will create an environment where employees are more likely to engage meaningfully, if they wish to, while respecting their autonomy.

9.3.4 Organisation-Wide Engagement

HR, health and safety teams and department managers should work together so that efforts are integrated into broader workplace policies and practices. A unified approach avoids fragmentation so that resources and messaging are consistent across the organisation. If you can integrate alcohol awareness into broader wellbeing strategies, this will highlight the interconnected nature of health issues. By embedding alcohol awareness into mental health programmes, stress management workshops and resilience training, you create a holistic approach to wellbeing that resonates with employees and addresses multiple aspects of their lives.

Now that we've covered the 'what' it's time to look at the 'how'.

9.4 DYNAMIC AWARENESS CAMPAIGNS

Awareness campaigns are most effective when they are engaging, relatable and accessible. By using visual elements such as infographics, video and illustrations, you can engage employees more effectively than with text-based communication alone. These visuals can simplify complex concepts and make the core message more memorable. For instance, graphics that demonstrate how alcohol affects the body or present statistics on alcohol consumption at work can reinforce the significance of the issue. Visuals are particularly beneficial in reaching diverse audiences, as they cater to different learning preferences, making the message accessible and engaging to all employees, regardless of their prior knowledge of the topic.

Along with visuals, the importance of concise and clear messaging cannot be overstated. In a workplace environment where employees are bombarded with a constant flow of information, short and to-the-point messages ensure that the key takeaways are retained. Also, your messages should motivate action, urging employees to access support services or get involved in wellness campaigns. Actionable messaging helps reinforce the idea that individuals have the power to make positive changes, whether it's by seeking help or adjusting their drinking habits.

Storytelling also plays a pivotal role in raising alcohol awareness. Sharing personal experiences or testimonials from employees who have faced alcohol-related challenges helps to humanise the issue. When employees hear relatable stories, it can reduce the stigma surrounding alcohol misuse. This approach not only strengthens engagement but encourages employees to feel more comfortable seeking and offering support. Stories provide a powerful platform for individuals to connect on an emotional level, making the issue less abstract and more tangible.

CASE STUDY

Inspired Outsourcing, an award-winning contact centre, produced a video for customers, urging them to take part in the 2018 Dry January campaign. Two employees shared their stories of losing a loved one to alcohol, whilst other team members held up a pledge sign and gave their reasons for going alcohol-free for a month. Finally, an employee completed a self-screening tool on camera and pledged his intention to improve his relationship with alcohol. This authentic, heartfelt message resonated with other businesses nationally.

For the message to reach as many employees as possible, you could use a variety of communication platforms. You might choose email newsletters to offer information on alcohol-related health risks and support resources, while digital signage and interactive posters can capture attention in common areas. For example, posters that include QR codes leading to wellness tools or educational materials can spark immediate engagement. Social media channels, whether internal or external, enable employees to share their experiences, or join initiatives like Dry January. By taking advantage of multiple streams, the message is more likely to resonate with employees across different touchpoints.

Alcohol Change UK has developed an excellent toolkit for framing conversations about alcohol.[7] Using the principles of choice architecture (that's Nudge Theory[8] to you and me), it explains how the manner in which we present ideas and information can influence the impact that they have. These messages shape how people are affected by and respond to information. The guide provides a set of evidence-based frames designed for communicating about alcohol and its associated harms. Research has demonstrated that these frames are effective in influencing public perceptions and opinions. The core message is that what we say about alcohol is only half the story, we need to think about how we impart information to engage and inspire the workforce.

CASE STUDY

A leading insurance broker held a 'pub quiz' alcohol awareness event, hosted by a stand-up comedian. Staff were encouraged to form their teams beforehand, choosing a name and designating a team captain. Non-alcoholic drinks and nibbles were provided the winning team received prizes. All questions were designed to challenge participants' basic knowledge of alcohol and debunk some common myths. Not only was the evening informative, but it also demonstrated that alcohol is not essential for an enjoyable night out.

9.5 CYCLE OF CHANGE

Where you have already identified that some employees are at risk of excessive drinking, you may wish to tailor activities to employees' different stages of readiness for change, as outlined in the Cycle of Change model (Chapter 4). Designing activities that are specific to these stages allows you to meet employees where they are in their behavioural change journey, offering the right level of support at the right time. An outline of the most effective tools for each stage of the process is given next.

9.5.1 Precontemplation Stage

At this stage, employees are not considering change and may be unaware of or in denial about the impact of alcohol on their health, work or relationships. The primary goal here is to increase awareness and begin the conversation about alcohol-related issues.

To engage employees at this stage, you could start by introducing educational materials like posters, videos and infographics that highlight the risks of alcohol misuse. The use of clear, non-judgemental language in these materials can open the door to further discussions. You could also offer anonymous screening, such as the use of the AUDIT tool online or handing out unit calculators.

Another effective strategy is to run informational webinars that address general alcohol misuse, along with the company's available support resources. These should aim to gently inform employees without pressuring them to change immediately. For example, an expert speaker could present the physiological effects of alcohol, how to work out units or drink-driving information and share how employees can seek help should they choose to do so later.

9.5.2 Contemplation Stage

In the contemplation stage, employees are aware of their alcohol consumption and may be thinking about reducing or changing their habits but are not yet committed to taking action. Activities for this stage should focus on encouraging deeper reflection about the potential consequences of drinking and motivating employees to start considering change. Sharing stories of employees who have successfully navigated alcohol-related challenges can help to make the issue more personal and relatable. Peer testimonials or case studies can provide a human element, helping others to see that change is not only possible but also beneficial.

Additionally, you could organise interactive workshops where employees discuss the personal, social and workplace impacts of alcohol use. This can facilitate group conversations, helping employees to express their concerns and receive feedback from others. By engaging employees in self-reflection and dialogue, you can create an environment that nurtures openness about alcohol and its potential to affect one's life.

9.5.3 Preparation Stage

When employees reach the preparation stage, they are ready to take action. They might be setting goals or preparing themselves mentally and physically to reduce their drinking. Here, the focus should be on providing the tools and support needed to start taking concrete steps.

A productive activity at this stage is helping employees set specific, realistic goals regarding their alcohol use. This could be done through workshops or one-to-one coaching, through an informal sober curious group or via delivery of brief advice (see Chapter 4) from your wellbeing champions. You should also gather information on the wide range of external resources that may be available to employees, such as counselling services, alcohol support hotlines or apps that track drinking habits.

9.5.4 Action Stage

Organising alcohol-free workplace events is a great way to encourage employees to participate in social activities without alcohol. These events can help employees socialise in a supportive environment and provide them with positive reinforcement that they can have fun without drinking.

Employees in the action stage can benefit from regular check-ins with managers or HR personnel. These meetings should provide a space for discussing challenges and reinforce the positive changes employees are making. Rewarding progress with incentives or recognition also helps employees feel supported and celebrated for their efforts.

Annual follow-ups or wellness check-ins help reinforce the focus on alcohol awareness and provide a platform for employees to assess their progress. You can continue to provide educational resources, highlight long-term benefits of alcohol reduction and remind employees of the support available if they ever face challenges.

9.6 AWARENESS-RAISING CAMPAIGNS

9.6.1 Dry January: A Month of Abstinence

Founded in 2013, the Dry January challenge is the most well-known alcohol abstinence campaign in the UK, challenging participants to abstain from alcohol for the entire month of January. The well-publicised campaign offers online and app-based support throughout the month and beyond and provides a structured opportunity for employees to reset and reduce their alcohol intake. Participants state that it encourages self-reflection on alcohol consumption habits and promotes personal discipline and self-awareness regarding alcohol use.

CAMPAIGN ACTIVITIES

- Launch the campaign with a company-wide announcement to generate enthusiasm and participation.
- Encourage employees to sign up officially on the Dry January website (this doubles the likelihood of completing the challenge).
- Offer incentives or rewards for employees who successfully complete Dry January.
- Share success stories and practical tips from employees who have participated in the past.
- Organise engaging non-alcoholic events, such as mocktail-making competitions or wellness workshops throughout the month.

Research suggests that around 8%[9] of those who successfully complete the month are still abstinent six months later, whilst other participants report financial savings, improved concentration, sleep and general health, weight loss and feeling more in control of their drinking.[10] From a health perspective, just thirty-one days alcohol-free can lead to improved insulin resistance and lowered blood pressure and reduce liver stiffness.[11]

CASE STUDY

A Midlands Health Authority launched a Dry January campaign amongst office staff, handing out Dry January flags for participants to place on their desk with the aim of encouraging participants to engage with and support each other.

Please note that this campaign is not for anyone who may be alcohol dependent. Participants are advised to undertake the AUDIT screening tool (see Chapter 4) and if their score is above the cut-off point of 19, to seek additional help before attempting a sobriety challenge.

9.6.2 Alcohol Awareness Week

Alcohol Awareness Week is an annual event, sponsored by Alcohol Change UK, that usually takes place in early July. Its aim is to raise awareness about alcohol-related issues, including health, social and workplace impacts. Alcohol Change UK provide free resources to anyone wishing to engage with the campaign. This is an excellent way of opening up conversations about alcohol and reducing stigma. Each campaign tackles a particular theme so year on year a range of relevant topics are covered. Recent topics include *Understanding Alcohol Harm*, *Alcohol and Cost* and *Alcohol and Mental Health.* Besides providing education on alcohol topics, the campaign encourages support for anyone facing alcohol-related challenges.

CAMPAIGN ACTIVITIES

- Host information sessions or workshops on the physical and mental health effects of alcohol.
- Share resources and helpline information for employees seeking assistance.
- Encourage employees to share their own stories or experiences related to alcohol, via blogs, vlogs and other media.
- Distribute informative materials and posters throughout the workplace to promote awareness.

9.6.3 Monthly Wellbeing Challenges

Implement a series of monthly wellbeing challenges that incorporate alcohol awareness into broader wellbeing themes. For example, focus on stress how alcohol affects stress levels in one month and explore the links with physical health harms in another. These can complement your wellbeing calendar by tying into Mental Health Awareness Week, Menopause Awareness Day and other wellbeing campaigns. This helps to maintain year-round awareness of alcohol's impact and strengthens the connection between alcohol consumption and specific health or emotional issues.

CAMPAIGN ACTIVITIES

- Develop a comprehensive schedule of monthly challenges aligned with broader wellbeing themes.
- Provide resources, articles and activities related to each month's focus.
- Create a digital platform or forum for employees to share their experiences, insights and progress.
- Recognise and reward employees who actively participate in the challenges.

9.6.4 Responsible Social Events

Ensure that all workplace social events promote responsible alcohol consumption practices and inclusivity. This will be important for employees who choose not to consume alcohol, with the added bonus of reducing the risk of alcohol-related incidents during social gatherings. By launching alcohol-aware events as a perk, you are tapping into the latest sober curious trends and taking the pressure off those who choose not to drink.

CAMPAIGN ACTIVITIES

- Collaborate with event organisers to provide a variety of non-alcoholic beverage options.
- Include alcohol awareness messages in event invitations and communications.
- Offer transportation alternatives to prevent drinking and driving after social events.

9.6.5 Topic-Specific Events

In previous chapters, we've established that for some populations such as women or people who are neurodivergent, there may be additional challenges. For example, the impact of alcohol on perimenopause and menopause is often poorly understood and women will often turn to alcohol to alleviate symptoms, unaware that drinking can have the opposite effect. One in three women say that they drink more since menopause and one in five women between the ages of forty-five and sixty-four regularly drink above recommended limits.[12] For people who have a diagnosis of ADHD or autism, a work environment tailored to neurotypicals can be difficult and distressing, so alcohol can become a way of fitting in or reducing (in the short-term) the impact of an overwhelming environment. The stigma that surrounds alcohol problems creates a further barrier to understanding that these responses are common. By introducing specific campaigns to raise awareness of these issues, employees may find it easier to recognise the difficulties that they may be having and seek more positive coping strategies.

For people from some cultural backgrounds, tackling excessive drinking brings additional barriers. There may be resources or charities that you could link to that offer support for specific minorities. For example, the excellent not-for-profit *No More Pretending* supports members of the Punjabi community to access help for themselves or loved ones who are struggling with alcohol.

9.7 ACTIVITIES TO RAISE AWARENESS

Finally, here are some additional ideas for activities to raise alcohol awareness.

TABLE 9.1
Table of Wellbeing Event Ideas to Raise Alcohol Awareness

Workshops	Invite experts to conduct workshops that explain the physical and psychological effects of alcohol, coping strategies and the importance of seeking help when needed. Using outside experts gives credibility to your campaigns and provides a different perspective on alcohol issues.
Health screenings	Organise holistic health check-up days that include assessments related to alcohol consumption, liver health and wellbeing. Screening results can be confidential and offer employees an opportunity to take positive action at an earlier stage than they otherwise might.
Health and wellbeing fairs	These can either target a range of wellbeing topics such as weight management, stress reduction, healthy eating and so forth or focus more specifically on alcohol. Activities could include informative workshops, health screenings, alcohol-free drink tastings, confidential support group sessions and informational booths. Additionally, your organisation could partner with a local addiction recovery service to provide accessible resources for employees seeking help.
Wellness challenges	Initiatives such as Dry January, Sober October and Alcohol Awareness Week have really gained momentum in the last few years. Promoting a campaign in the workplace can encourage employees to take part and provide additional support to help them to achieve their aims.
Sober curious groups	In-house support groups have proven effective for a number of topics and the growth of sober curious or abstinence-based social groups is no exception. Arranging sober events or just linking up with like-minded colleagues can be a positive perk of the workplace.
Confidential support groups	Create a safe space for employees to share experiences, challenges and strategies for coping with alcohol-related issues. Regular meetings can take place on-line for larger organisations and these can be of real benefit to those working towards or maintaining sobriety.
Guest Speakers	Invite individuals in recovery from alcohol dependence or experts in addiction psychology to share their personal stories and professional insights. Often, these glimpses into the difficulties that people with an AUD face resonate with employees who may be struggling with their own drinking or who are supporting a loved one. Ensure that routes to additional support are offered to personal alongside such events.
Charity partnerships	Whilst raising awareness of alcohol, you can also contribute to reducing harm. Consider choosing an alcohol charity for a fundraising partnership. Not only is this an excellent conversation starter but also helps out for a good cause too. Some of the national charities that work towards harm reduction, education and support are listed here, although you may know of other relevant local charities. National alcohol-related charities include: Adfam, Alcohol and Families Alliance, Alcohol Change UK, Alcohol Focus Scotland, The Anti-stigma network, NACOA, National FASD.

TABLE 9.1 (*Continued*)
Table of Wellbeing Event Ideas to Raise Alcohol Awareness

Webinars and seminars	Lunch and Learn initiatives are great for providing bitesize introductions to the topic. They can focus on any aspect of alcohol awareness that is relevant to the workforce from Understanding Alcohol Units and Limits to How to Support a Loved one with an Alcohol Addiction.
Websites and other online tools	For larger organisations, consider the development of a website that contains alcohol awareness information, the AUDIT screening tool, blogs and success stories, promotes wellbeing activities and signposts employees to your policy and to additional support. This would provide a valuable resource for employees and can be easily monitored and maintained. Other online tools such as closed Facebook pages for employee mutual support, a Q&A portal or internet-based recovery options, can also be developed.

NOTES

1 British Medical Association, 2015. *Alcohol, Drugs and the Workplace: The Role of Medical Professionals*. London: BMA (https://www.bma.org.uk/media/1067/bma_alcohol-and-drugs-in-the-workplace-_oct_2019.pdf).

2 Public Health England, 2016. *The Public Health Burden of Alcohol and the Effectiveness and Cost-Effectiveness of Alcohol Control Policies*. London: PHE (https://www.gov.uk/government/publications/the-public-health-burden-of-alcohol-evidence-review).

3 International Labour Organization, 2021. *Workplace Health Promotion and Wellbeing*. Geneva: ILO, (Blog) (https://www.ilo.org/workplace-health-promotion-and-well-being).

4 Office for Health Improvement & Disparities, 2023. UK Clinical Guidelines for Alcohol Treatment: Specific Settings and Populations (https://www.gov.uk/government/consultations/uk-clinical-guidelines-for-alcohol-treatment/uk-clinical-guidelines-for-alcohol-treatment-specific-settings-and-populations#residential-treatment-and-intensive-structured-day-treatment).

5 Pidd, et al., 2018. Workplace Alcohol Harm Education Intervention in Australia: Cluster Non-Randomised Controlled Trial. *Drug and Alcohol Review* May; 37(4), 502–513.

* Chapter five.

6 Rosenberg, et al., 2018. New National Alcohol Guidelines in the UK: Public Awareness, Understanding and Behavioural Intentions. *Journal of Public Health* Sep 1; 40(3), 549–556.

7 Alcohol Change UK, 2024. How to Talk about Alcohol (https://s3.eu-west-2.amazonaws.com/sr-acuk-craft/documents/Alcohol-Toolkit_Final.pdf).

8 Thaler, R. H. and Sunstein, C. R., 2008. *Nudge: Improving Decisions about Health, Wealth and Happiness*. New Haven: Yale University Press.

9 De Visser, et al., 2016. Voluntary Temporary Abstinence from Alcohol During Dry January and Subsequent Alcohol Use (https://pubmed.ncbi.nlm.nih.gov/26690637/).

10 de Visser, R. O., 2019. *Evaluation of Dry January 2019*. School of Psychology, University of Sussex (https://s3.eu-west-2.amazonaws.com/sr-acuk-craft/documents/R.-de-Visser-Dry-January-evaluation-2019.pdf).

11 Mehta, et al., 2018. Short-Term Abstinence from Alcohol and Changes in Cardiovascular Risk Factors, Liver Function Tests and Cancer-Related Growth Factors: A Prospective Observational Study. *BMJ Open* (https://pmc.ncbi.nlm.nih.gov/articles/PMC5942469/#:~:text=The%20key%20findings%20of%20this,month%20of%20abstinence%20from%20alcohol).
12 NHS Digital Health Survey for England 2021 (https://digital.nhs.uk/data-and-information/publications/statistical/health-survey-for-england/2021/part-3-drinking-alcohol#:~:text=of%20pure%20alcohol.-,Frequency%20of%20drinking%20in%20the%20last%2012%20months%2C%20by%20age,to%20do%20so%20(59%25)).

10 Measuring Effectiveness and Moving Forward

10.1 INTRODUCTION

Over the last nine chapters, we've explored the role of alcohol in UK society, its history and its significance and its potential for harm. We've explored how the workplace can play a pivotal role in promoting responsible consumption and reducing those harms, whilst improving business outcomes. Having launched your policy, set up employee support options and integrated alcohol awareness into your wellbeing strategy, you'll want to know how well your hard work is paying off.

A good evaluation of the effectiveness of your policies and practices will allow you to refine your initiative and enhance outcomes in the longer term. A systematic approach that integrates data collection, stakeholder engagement, financial evaluation and continuous improvement not only justifies the allocation of resources but also helps identify areas for improvement.

10.2 THE IMPORTANCE OF MEASURING EFFECTIVENESS

Firstly, evaluation reinforces organisational accountability by ensuring that businesses uphold their commitments to safeguarding employee wellbeing. By tracking outcomes, you can show that the organisation is actively managing these issues and adhering to its stated goals. For example, publishing results in internal reports can show leadership's commitment and transparency.

Secondly, continuous improvement relies on regular assessments to highlight which aspects of the interventions are effective and where adjustments are needed. This ongoing evaluation allows you to tweak your activities based on real-world results, so that your input remains relevant and impactful. Without regular feedback loops and performance tracking, even well-designed programmes risk becoming outdated or ineffective in changing contexts.

Thirdly, effective measurement allows you to allocate resources effectively. By analysing which interventions generate the greatest impact relative to cost, you can prioritise funding for those initiatives that offer the best ROI. For instance, if data shows that peer support projects lead to substantial reductions in alcohol misuse, resources can be shifted from less effective campaigns to scale up these successful efforts.

As always, you should engage employees in the evaluation process. When workers see evidence that interventions are making a difference—such as improvements in workplace culture or access to support services—they are more likely to participate enthusiastically and advocate for these programmes.

Finally, monitoring helps to establish that alcohol-related interventions adhere to regulations, such as data protection laws and health and safety standards. For

DOI: 10.1201/9781032648927-11

example, ensuring that employee data collected through surveys remains confidential and is used appropriately is both a legal requirement and an ethical responsibility. Evaluations can also reveal any potential risks or areas where policies need to be adjusted to remain compliant with changing legal requirements.

10.2.1 Understanding the Goals of Alcohol-Related Interventions

Before considering evaluation techniques, you will have defined the specific goals of your intervention. Goals have been discussed in previous chapters and need to be tailored to your workforce and it's essential to keep sight of your objectives to guide the selection of appropriate evaluation methods.

10.2.2 Establishing Baseline Data

The first step is to collect baseline data. This provides a point of comparison for assessing changes over time. Baseline data can be gathered through surveys, health screenings or workplace assessments, depending on the intervention's focus. For example, you may want to include rates of absenteeism, productivity statistics or employee self-reports of alcohol use. If you are not already collecting these data regularly, this is a great opportunity to implement and standardise data collection methods. This will stand you in good stead going forward by improving the validity and reliability of results. Collecting robust baseline data will mean that subsequent evaluations can accurately attribute changes to the intervention rather than external factors.

10.3 KEY METRICS FOR MEASURING EFFECTIVENESS

You can use a variety of metrics to evaluate the effectiveness of your interventions. Consider your aims. Well-defined goals will be measurable. You may wish to include both outcome measures (assessing changes in behaviour or absenteeism) and process measures (evaluating the implementation and reach of the intervention). In Chapter 5, you were asked to reflect on the financial costs of absenteeism and presenteeism and gather baseline data. In Chapter 9, the question *what is the primary aim of your alcohol awareness initiatives?* was posed. So, if you haven't already done so, establishing your aims and how you will recognise or measure success will be your next step. Next are some of the key data you may wish to capture.

10.3.1 Reduction in Alcohol Consumption

This is often a primary indicator of success, although identifying a measurable reduction in alcohol intake among employees is challenging. One way to measure reduction is to track self-reported surveys or tools such as AUDIT scores (see Chapter 4). You could also determine how many employees participate in Dry January (Chapter 9) or other sobriety or alcohol reduction campaigns. Be aware though, that reduction in consumption doesn't always equate to a reduction in harm or risk. Employees drinking moderately are those who are most likely to engage with your alcohol-related wellbeing initiatives. For larger organisations, measuring the take up of offers of support from EAPs or counselling would be a useful way of assessing reductions in harm.

10.3.2 Improved Health Outcomes

Health metric gains such as improved sleep, more energy, greater concentration and improvements in mental health or stress levels can provide subjective evidence of intervention success. Self-reported data on improvements can have a positive impact on engagement and encourage increased participation. For employees who use wearable fitness trackers or have access to a company-sponsored health scheme, data such as improved heart rate variability or resting heart rate can provide objective, measurable results. Combining both self-reporting and objective measurements increases the reliability of your data.

10.3.3 Workplace Metrics

From an occupational perspective, key metrics include reduced absenteeism, fewer alcohol-related incidents and improvements in productivity. For more qualitative data, employee engagement surveys can provide insights into perceptions of workplace responses around alcohol. You could also look at data on accidents and near misses and compare your findings to other comparable organisations. Monitoring adherence to workplace alcohol policies can be used to measure compliance and deterrence. If you have introduced alcohol testing, you will want to examine how many non-negative tests results are performed and in what circumstances. You can then determine what action needs to be taken to reduce these numbers.

10.3.4 Knowledge and Awareness

Pre- and post-intervention surveys assessing participants' understanding of alcohol risks and harm can gauge educational impact. Increased awareness often correlates with behaviour change over time. For best effect, educational input should be repeated at regular intervals, such as biennially. You can then measure the cumulative effect of your messaging too.

10.3.5 Engagement and Participation Rates

Tracking participation rates in workshops, seminars and other intervention activities can indicate programme reach and relevance. Higher engagement often reflects greater acceptance and potential impact. It can be useful to examine whether participation rates vary between different departments of the workforce or at different times of the year. For example, campaigns are often targeted at just before the summer break or in the run-up to the festive season. It's useful to evaluate your messaging around alcohol. Do some platforms or messages achieve a greater response than others? Are your communications reaching their intended audience? Are your messages clear and easily understood by your workforce?

10.3.6 Selecting Appropriate Evaluation Methods

Effective evaluation requires a combination of quantitative and qualitative methods. Quantitative methods, such as surveys and health screenings, provide measurable data on outcomes like alcohol consumption, health status and workplace performance. You can also use standardised tools such as the AUDIT to assess drinking behaviours. Changes in AUDIT scores over time can indicate a reduction in harm and an increase in employee productivity, as a result of changed alcohol consumption.

Qualitative methods, including interviews and focus groups, offer valuable insights into participants' experiences and thoughts. These methods are particularly useful for understanding the contextual factors that influence behaviour change, such as cultural attitudes towards alcohol and barriers to accessing support services. Methods should always be tailored to the context of the intervention.

10.4 DATA COLLECTION METHODS

10.4.1 Surveys and Questionnaires

These are commonly used tools for gathering quantitative data on employee awareness, participation and attitudes towards alcohol initiatives. You can design structured surveys to capture changes in knowledge and behaviour before and after interventions. For example, using the AUDIT screening tool provides standardised measures to evaluate risk levels across the workforce. In addition to surveys, focus groups and interviews offer qualitative insights that delve deeper into personal experiences and perceptions. These methods provide nuanced understanding by capturing emotional and psychological responses, which are often overlooked in quantitative analyses.

10.4.2 Interviews and Focus Groups

Interviews with employees can reveal individual challenges and successes, while focus groups facilitate open discussions that highlight collective attitudes and barriers. Qualitative data collected through focus groups and interviews offer rich insights into employees' thoughts, attitudes and personal experiences with alcohol-related initiatives. These methods provide a deeper understanding of the emotional and psychological aspects of your intervention's impact.

10.4.3 Behavioural Observation

By monitoring participation in workshops, adherence to policies and attendance at events, you can collect tangible evidence of behaviour change. Observational data, when combined with self-reported information, paints a comprehensive picture of intervention outcomes. However, these more qualitative data are most reliable when coupled with quantitative data collection methods. You could document individual or organisational case studies to highlight successes and challenges, offering practical lessons for future interventions. This is particularly useful when an employee has successfully completed an alcohol treatment regime. They may be willing to reflect on what was helpful and what challenges they faced at work.

10.4.4 Technology-Based Tracking

This enhances data collection by using digital tools such as wellness apps and wearable devices. These platforms offer real-time monitoring of alcohol-related behaviours, enabling personalised feedback and more dynamic evaluation processes. Self-reporting through these tools allows for greater flexibility, though it requires honesty and trust among participants. Routine monitoring of health records, absenteeism data and productivity records can provide objective evidence of intervention impact. Data analytics tools can help identify trends and correlations. Consider that different departments may show different results, giving you useful information on where to tailor or expand your efforts.

10.4.5 Randomised Controlled Trials

For large-scale interventions, randomised controlled trials (RCTs) can provide high-quality evidence of effectiveness. However, these require costly resources and often involve complex ethical considerations so are unlikely to be conducted in most organisations. Occasionally though, an opportunity to participate in academic research may come up. Research findings can provide excellent information that can guide your future activities.

10.4.6 Monitoring Progress

Once you've gathered baseline data you should be follow up with periodic evaluations at set intervals, such as six months or one year, to track the longevity of any improvements. If you've identified key performance indicators, you'll have a valuable aid in tracking progress. To monitor effectively, you'll need to repeat your regular data collection, hold periodic reviews and make iterative adjustments based on findings. Providing anonymous feedback conduits can encourage honest input, thereby highlighting both the strengths of the intervention and areas that may require adjustment or enhancement.

Continuous monitoring can be time-consuming and resource-intensive but is justified to ascertain the longer-term impact of your policy and practices. It will provide valuable insights into the sustainability of behaviour change and the factors that support or hinder progress. For example, a reduction in alcohol-related absenteeism can be a strong indicator of success but the results won't be visible immediately, so it makes sense to monitor trends over several years. There may be additional unforeseen benefits such as improved work outputs or potential challenges such as higher staff turnover in response to a new testing regime that only become apparent when you start to study the data. To facilitate the ongoing monitoring process, you may wish to invest in data management systems that allow for efficient information collection, storage and analysis.

10.4.7 Data-Driven Adjustments

Interventions should be flexible and responsive to feedback and emerging evidence. Evaluation, therefore, should not be a one-time exercise but rather an ongoing process that responds to changes and challenges. Regular evaluation reports can inform decision-making and policy or intervention adjustments. For example, if data reveals low participation rates, you might revise your communication strategy to increase awareness and engagement. If participants report barriers to accessing support services, efforts can be made to address these barriers, such as by offering services in more convenient locations or providing digital options.

In some cases, evaluation findings may suggest the need for more fundamental changes to the intervention's design. For example, a workplace may decide to implement a more comprehensive alcohol policy that includes regular training for managers and employees on recognising and addressing alcohol-related issues. Conversely, interventions that demonstrate greater impact can be scaled or adapted for broader implementation. Don't forget to share good practice and lessons learned as well as celebrating achievements.

If you can engage stakeholders in the evaluation process, so much the better as it will ensure that their perspectives are considered and that interventions are tailored to meet their needs. Regular consultation with staff, managers, senior leadership and potentially customers, improves buy-in for your strategies and can lead to greater adherence to policies.

10.4.8 Reporting and Disseminating Findings

The final step in the evaluation process is to report and disseminate the findings. Clear reporting will give stakeholders confidence in the intervention's outcomes and the implications for future practice. Reports should include an overview of the intervention, the methods used for evaluation, the findings and suggestions for improvement. Disseminating findings can take various forms, including published reports, presentations and workshops with stakeholders. Your reporting may be predominantly statistical in nature, focusing on financial savings or reductions in uninsured losses, for example or you may wish to report health gains by using case studies. But remember to tailor your dissemination method to your audience: Dry January participants probably don't need to be informed about a 2% reduction in absenteeism during January but would love to hear that their sponsorship has raised funds for a good cause.

10.5 LONG-TERM EVALUATION AND SUSTAINABILITY

While short-term results provide valuable insights, long-term evaluation is essential for understanding the sustained impact of your interventions. Tracking outcomes over several years allows you to measure changes in absenteeism and productivity which are effective indicators of long-term success. Longitudinal evaluations help identify patterns and trends that may not be apparent in immediate post-intervention assessments. If you work in a larger organisation, you could call in external consultants who can design a bespoke monitoring cycle, crunch the data and provide you with a slick report of the results, which may involve an initial financial commitment but could save time and money in the long run.

10.5.1 Financial Impact: Cost-Benefit Analysis and Return on Investment

It's important to evaluate the financial impact of your programme to justify investments and guide future decisions. Conducting a cost-benefit analysis, by comparing the expenses associated with implementing these initiatives against the benefits they generate, will help you to establish whether your interventions are cost-effective. You may wish to include assessing reductions in absenteeism or non-negative alcohol tests, improvements in productivity and decreases in healthcare costs. But your analysis can also include more intangible benefits such as improved employee well-being, stronger workplace relationships and reduced conflict.

Calculating and expressing the ROI as a percentage offers a straightforward, quantifiable measure of an initiative's financial success. This approach helps you to understand whether the benefits gained from your alcohol-focused intervention outweigh the costs associated with its implementation. For example, a company might invest in training sessions, wellness or awareness campaigns and then track related outcomes such as reduced absenteeism or improved productivity. By converting the net financial benefit into a percentage relative to the initial investment, decision-makers gain a clear, comparable metric that highlights the initiative's effectiveness. This clarity allows leaders to assess whether the intervention adds value and can influence decisions about future resource allocation.

CALCULATING ROI

Use the formula:
ROI = (net benefit) (cost of intervention) x 100

Net benefit = benefits (e.g. reduced absenteeism, increased productivity)—costs (e.g. programme expenses, staff time)

10.5.2 Benchmarking and Adopting Best Practices

This is another way to examine effectiveness. By benchmarking your results against industry standards and comparing outcomes with those of similar organisations, you can identify areas where you excel or need improvement. You can research successful strategies and adapt them to fit the context of your workplace: why reinvent the wheel? Where others in your industry have been successful, it's likely that you may also have a measure of success. The sector is continually evolving, and organisations that keep pace with new developments are better positioned to implement cutting-edge interventions, so you need to say informed about emerging trends and technologies. Benchmarking efforts should include a review of **KPIs** that correspond with industry norms (or could establish industry norms), ensuring that your evaluations remain relevant. It is only over the last few years that businesses have really started to see the value in tackling alcohol harm, so there is still a relative dearth of examples of effective workplace programmes. This means that your work in this area can also provide an important contribution to the field, going forward. *You* might be the innovator who lights the way for your sector.

10.5.3 Risk Assessments and Crisis Management

This is perhaps the more important aspect of evaluation. Risk assessments play a vital role in identifying and mitigating potential problems related to alcohol use. Baseline and ongoing data should inform your risk management strategy. Your data analysis should be able to pinpoint high-risk areas and employee groups. This proactive approach lends itself to developing targeted interventions that address specific vulnerabilities. Consulting with experts (in both health and safety and alcohol in the workplace) will enhance the effectiveness of your risk mitigation strategies. Alongside this, you can develop protocols for crisis response, safeguarding the organisation in the event of alcohol-related incidents.

10.6 MOVING FORWARD

As we conclude our journey through the subject of alcohol use in the workplace, it is time to reflect on the knowledge and insights gained from the preceding chapters. We have explored the prevalence of alcohol misuse, its impact on employees and organisations, the importance of early intervention and best practice for prevention and support. Now, let's look to the future, where the call to action is clear: it is time to move forward together in addressing alcohol problems in the workplace.

The rationale for action is undeniable. Alcohol misuse in the workplace is not just a hidden epidemic; it is a pressing concern that affects employees, their families, communities and the fabric of our organisations. From the financial toll it takes to the emotional burdens carried by individuals and their loved ones, the consequences of inaction are boundless.

Future research and practices in addressing alcohol-related challenges are evolving, with several emerging opportunities to enhance effectiveness and inclusivity. One key area is the development of personalised alcohol interventions, designed to meet the unique needs of individuals based on their circumstances. These tailored strategies could focus on factors such as an employee's risk level, work environment or personal stressors, increasing the likelihood of meaningful behaviour change. For instance, an employee facing high job-related stress may benefit from stress management techniques integrated into alcohol awareness programmes.

The use of technological solutions also holds promise. Advanced tools, such as AI-driven platforms, can enable real-time monitoring of alcohol consumption patterns and provide early interventions. Mobile applications designed to support wellbeing may offer personalised recommendations, reminders or access to professional support services; digital data collection can streamline the evaluation process, offering insights that were previously difficult to obtain.

Considering neurodiversity and gender in future interventions is important, too. Alcohol misuse may present differently in neurodivergent individuals, necessitating adjustments in communication, support and intervention design. Similarly, gender-specific approaches can address the distinct social, psychological and biological factors that influence alcohol-related behaviours among different groups. For example, women may face unique barriers to seeking help due to societal stigma, while men might be more influenced by workplace culture.

In a global context, understanding cultural variations in alcohol use and the effectiveness of interventions is also highly relevant. What works in one cultural setting may not translate directly to another. If you have an international workforce, your research should focus on adapting strategies to different norms, beliefs and values, ensuring that your interventions are appropriate and relevant.

Lastly, long-term impact studies are necessary to evaluate the sustainability of alcohol awareness initiatives. Short-term success does not always equate to lasting change, so longitudinal research can help organisations refine their approaches. Tracking outcomes over multiple years will provide valuable data on whether initial gains are maintained and highlight areas that require ongoing attention or modification.

By embracing these forward-looking strategies, we can create more inclusive and sustainable approaches to tackling alcohol-related issues, ultimately promoting a healthier and more productive environment.

We also need to champion the successes and positive changes that result from alcohol-awareness initiatives. Recognise and reward individuals or teams that make strides in their wellbeing journey.

In conclusion, I invite you to take these lessons to heart, to engage in open dialogue and to work collaboratively towards a future where alcohol misuse in the workplace is no longer hidden but openly addressed and effectively managed. Employers and employees must come together to promote responsible alcohol use, support those in need and ultimately sustain healthier and more productive workplaces. Let's not keep this knowledge to ourselves. Share your insights with peers, colleagues, family and friends. Working together, we can initiate worthwhile change that improves the wellbeing and resilience of all employees. Remember that change starts with awareness, compassion and action and by implementing the strategies outlined in this book, you can make an important difference in the lives of the people you work with.

FINAL QUESTIONS FOR REFLECTION

What's the most important thing you've learned from reading this book?

Consider not only strategies for reducing workplace harm but also reflect on whether your reading has affected your thoughts about alcohol and drinking culture on a personal level.

How are you going to turn your knowledge into action?

Glossary

Alcohol Use Disorder (AUD) A pattern of alcohol consumption that causes psychological or physical health problems, often characterised by the inability to control alcohol use despite harmful consequences.

Alcohol Use Disorders Identification Test (AUDIT) A validated screening tool to determine whether an individual's drinking may be putting them at risk of health harms. This tool is not cannot be used for diagnosing dependence.

Blood Alcohol Concentration (BAC) A measurement of the amount of alcohol in a person's bloodstream, typically used to assess intoxication levels.

Cognitive Behavioural Therapy (CBT) A talk therapy that challenges thoughts, feelings and beliefs to help people to change behaviours.

Employee Assistance Programme (EAP) A confidential workplace scheme offering support and resources to employees dealing with personal issues, including those related to alcohol.

General Data Protection Regulation (GDPR) Guidance detailing data and privacy law

Key Performance Indicator (KPI) A set of measurable criteria for success of a project or venture.

National Institute for Health and Care Excellence (NICE) Sector of the department of Health and Social care that publishes guidance on health and clinical issues.

Standard Alcohol Unit (SAU) A standard measure used to quantify the amount of pure alcohol in a drink, with one unit equating to 8 g or 10 ml of pure ethanol.

Index

FSC
www.fsc.org
MIX
Papier | Fördert
gute Waldnutzung
FSC® C083411

Kolibri 360 GmbH
Ferdinand-Jühlke-Straße 7
99095 Erfurt, Deutschland
produktsicherheit@kolibri360.de